ISBN 0-8373-0734-1

C-734 CAREER EXAMINATION SERIES

This is your PASSBOOK® for...

Sewage Treatment Worker

Test Preparation Study Guide

Questions & Answers

NLC

NATIONAL LEARNING CORPORATION

National Learning Corporation

212 Michael Drive, Syosset, NY 11791
(516) 921-8888 • www.passbooks.com
E-mail: info@passbooks.com

PUBLISHED IN THE UNITED STATES OF AMERICA

PASSBOOK® SERIES

THE *PASSBOOK® SERIES* has been created to prepare applicants and candidates for the ultimate academic battlefield – the examination room.

At some time in our lives, each and every one of us may be required to take an examination – for validation, matriculation, admission, qualification, registration, certification, or licensure.

Based on the assumption that every applicant or candidate has met the basic formal educational standards, has taken the required number of courses, and read the necessary texts, the *PASSBOOK® SERIES* furnishes the one special preparation which may assure passing with confidence, instead of failing with insecurity. Examination questions – together with answers – are furnished as the basic vehicle for study so that the mysteries of the examination and its compounding difficulties may be eliminated or diminished by a sure method.

This book is meant to help you pass your examination provided that you qualify and are serious in your objective.

The entire field is reviewed through the huge store of content information which is succinctly presented through a provocative and challenging approach – the question-and-answer method.

A climate of success is established by furnishing the correct answers at the end of each test.

You soon learn to recognize types of questions, forms of questions, and patterns of questioning. You may even begin to anticipate expected outcomes.

You perceive that many questions are repeated or adapted so that you can gain acute insights, which may enable you to score many sure points.

You learn how to confront new questions, or types of questions, and to attack them confidently and work out the correct answers.

You note objectives and emphases, and recognize pitfalls and dangers, so that you may make positive educational adjustments.

Moreover, you are kept fully informed in relation to new concepts, methods, practices, and directions in the field.

You discover that you are actually taking the examination all the time: you are preparing for the examination by "taking" an examination, not by reading extraneous and/or supererogatory textbooks.

In short, this PASSBOOK®, used directedly, should be an important factor in helping you to pass your test.

SEWAGE TREATMENT WORKER

DUTIES
Sewage Treatment Workers under general supervision, operate, maintain and repair machinery, equipment and structures in wastewater treatment plants, clean water storage and processing facilities, pumping stations, intercepting sewers, sludge treatment and disposal facilities and associated equipment and facilities; test water and sewage samples and perform confined space entry duties. They operate, maintain and repair equipment such as regulators, diversion chambers, tide gates, pumps, sludge collecting treatment and disposal equipment, blowers, compressors, motors, air filters, oil purifying equipment, chlorination equipment, heat exchanger equipment, boilers, valves, pipes and meters in wastewater treatment plants, clean water storage and processing facilities, pumping stations, intercepting sewers, sludge treatment and disposal facilities, and related facilities; take samples and perform tests relating to sewage composition and clean water quality; participate in the operation and monitoring of chemical bulk storage tanks, such as making periodic inspections and tests as required, reporting tank levels, monitoring leakage warning systems and deployment of spill containment devices; read meters, gauges and charts and records the readings; keep operating logs and other records; prepare reports and requisitions; serve as confmed space entry attendant; clean and maintain structures, equipment and grounds; load sludge vessels and barges; operate vehicles and radio telephones; clean, wash, and perform minor maintenance on vehicles; perform rigging and hoisting; handle grit, screenings, grease and sludge; operate welding and burning equipment; operate and maintain power tools;

SCOPE OF THE EXAMINATION
The multiple-choice test may include questions on basic principles of the operation, maintenance, installation and repair of mechanical equipment such as pumps, valves, hydraulic equipment, conveyors, heat exchangers, compressors, and heating, ventilation and air conditioning equipment; basic principles of pipefitting, burning and welding, carpentry, masonry, painting, hoisting and rigging, locksmithing and roofing; maintenance and safe operation of power and hand tools and general safety; understanding of written instructions and drawings; arithmetic; and other related areas.

HOW TO TAKE A TEST

I. YOU MUST PASS AN EXAMINATION

A. WHAT EVERY CANDIDATE SHOULD KNOW

Examination applicants often ask us for help in preparing for the written test. What can I study in advance? What kinds of questions will be asked? How will the test be given? How will the papers be graded?

As an applicant for a civil service examination, you may be wondering about some of these things. Our purpose here is to suggest effective methods of advance study and to describe civil service examinations.

Your chances for success on this examination can be increased if you know how to prepare. Those "pre-examination jitters" can be reduced if you know what to expect. You can even experience an adventure in good citizenship if you know why civil service exams are given.

B. WHY ARE CIVIL SERVICE EXAMINATIONS GIVEN?

Civil service examinations are important to you in two ways. As a citizen, you want public jobs filled by employees who know how to do their work. As a job seeker, you want a fair chance to compete for that job on an equal footing with other candidates. The best-known means of accomplishing this two-fold goal is the competitive examination.

Exams are widely publicized throughout the nation. They may be administered for jobs in federal, state, city, municipal, town or village governments or agencies.

Any citizen may apply, with some limitations, such as the age or residence of applicants. Your experience and education may be reviewed to see whether you meet the requirements for the particular examination. When these requirements exist, they are reasonable and applied consistently to all applicants. Thus, a competitive examination may cause you some uneasiness now, but it is your privilege and safeguard.

C. HOW ARE CIVIL SERVICE EXAMS DEVELOPED?

Examinations are carefully written by trained technicians who are specialists in the field known as "psychological measurement," in consultation with recognized authorities in the field of work that the test will cover. These experts recommend the subject matter areas or skills to be tested; only those knowledges or skills important to your success on the job are included. The most reliable books and source materials available are used as references. Together, the experts and technicians judge the difficulty level of the questions.

Test technicians know how to phrase questions so that the problem is clearly stated. Their ethics do not permit "trick" or "catch" questions. Questions may have been tried out on sample groups, or subjected to statistical analysis, to determine their usefulness.

Written tests are often used in combination with performance tests, ratings of training and experience, and oral interviews. All of these measures combine to form the best-known means of finding the right person for the right job.

II. HOW TO PASS THE WRITTEN TEST

A. NATURE OF THE EXAMINATION

To prepare intelligently for civil service examinations, you should know how they differ from school examinations you have taken. In school you were assigned certain definite pages to read or subjects to cover. The examination questions were quite detailed and usually emphasized memory. Civil service exams, on the other hand, try to discover your present ability to perform the duties of a position, plus your potentiality to learn these duties. In other words, a civil service exam attempts to predict how successful you will be. Questions cover such a broad area that they cannot be as minute and detailed as school exam questions.

In the public service similar kinds of work, or positions, are grouped together in one "class." This process is known as *position-classification*. All the positions in a class are paid according to the salary range for that class. One class title covers all of these positions, and they are all tested by the same examination.

B. FOUR BASIC STEPS

1) Study the announcement

How, then, can you know what subjects to study? Our best answer is: "Learn as much as possible about the class of positions for which you've applied." The exam will test the knowledge, skills and abilities needed to do the work.

Your most valuable source of information about the position you want is the official exam announcement. This announcement lists the training and experience qualifications. Check these standards and apply only if you come reasonably close to meeting them.

The brief description of the position in the examination announcement offers some clues to the subjects which will be tested. Think about the job itself. Review the duties in your mind. Can you perform them, or are there some in which you are rusty? Fill in the blank spots in your preparation.

Many jurisdictions preview the written test in the exam announcement by including a section called "Knowledge and Abilities Required," "Scope of the Examination," or some similar heading. Here you will find out specifically what fields will be tested.

2) Review your own background

Once you learn in general what the position is all about, and what you need to know to do the work, ask yourself which subjects you already know fairly well and which need improvement. You may wonder whether to concentrate on improving your strong areas or on building some background in your fields of weakness. When the announcement has specified "some knowledge" or "considerable knowledge," or has used adjectives like "beginning principles of..." or "advanced ... methods," you can get a clue as to the number and difficulty of questions to be asked in any given field. More questions, and hence broader coverage, would be included for those subjects which are more important in the work. Now weigh your strengths and weaknesses against the job requirements and prepare accordingly.

3) Determine the level of the position

Another way to tell how intensively you should prepare is to understand the level of the job for which you are applying. Is it the entering level? In other words, is this the position in which beginners in a field of work are hired? Or is it an intermediate or advanced level? Sometimes this is indicated by such words as "Junior" or "Senior" in the class title. Other jurisdictions use Roman numerals to designate the level – Clerk I, Clerk II, for example. The word "Supervisor" sometimes appears in the title. If the level is not indicated by the title, check the description of duties. Will you be working under very close supervision, or will you have responsibility for independent decisions in this work?

4) Choose appropriate study materials

Now that you know the subjects to be examined and the relative amount of each subject to be covered, you can choose suitable study materials. For beginning level jobs, or even advanced ones, if you have a pronounced weakness in some aspect of your training, read a modern, standard textbook in that field. Be sure it is up to date and has general coverage. Such books are normally available at your library, and the librarian will be glad to help you locate one. For entry-level positions, questions of appropriate difficulty are chosen – neither highly advanced questions, nor those too simple. Such questions require careful thought but not advanced training.

If the position for which you are applying is technical or advanced, you will read more advanced, specialized material. If you are already familiar with the basic principles of your field, elementary textbooks would waste your time. Concentrate on advanced textbooks and technical periodicals. Think through the concepts and review difficult problems in your field.

These are all general sources. You can get more ideas on your own initiative, following these leads. For example, training manuals and publications of the government agency which employs workers in your field can be useful, particularly for technical and professional positions. A letter or visit to the government department involved may result in more specific study suggestions, and certainly will provide you with a more definite idea of the exact nature of the position you are seeking.

III. KINDS OF TESTS

Tests are used for purposes other than measuring knowledge and ability to perform specified duties. For some positions, it is equally important to test ability to make adjustments to new situations or to profit from training. In others, basic mental abilities not dependent on information are essential. Questions which test these things may not appear as pertinent to the duties of the position as those which test for knowledge and information. Yet they are often highly important parts of a fair examination. For very general questions, it is almost impossible to help you direct your study efforts. What we can do is to point out some of the more common of these general abilities needed in public service positions and describe some typical questions.

1) General information

Broad, general information has been found useful for predicting job success in some kinds of work. This is tested in a variety of ways, from vocabulary lists to questions about current events. Basic background in some field of work, such as sociology or economics, may be sampled in a group of questions. Often these are principles which have become familiar to most persons through exposure rather than through formal training. It is difficult to advise you how to study for these questions; being alert to the world around you is our best suggestion.

2) Verbal ability

An example of an ability needed in many positions is verbal or language ability. Verbal ability is, in brief, the ability to use and understand words. Vocabulary and grammar tests are typical measures of this ability. Reading comprehension or paragraph interpretation questions are common in many kinds of civil service tests. You are given a paragraph of written material and asked to find its central meaning.

3) Numerical ability

Number skills can be tested by the familiar arithmetic problem, by checking paired lists of numbers to see which are alike and which are different, or by interpreting charts and graphs. In the latter test, a graph may be printed in the test booklet which you are asked to use as the basis for answering questions.

4) Observation

A popular test for law-enforcement positions is the observation test. A picture is shown to you for several minutes, then taken away. Questions about the picture test your ability to observe both details and larger elements.

5) Following directions

In many positions in the public service, the employee must be able to carry out written instructions dependably and accurately. You may be given a chart with several columns, each column listing a variety of information. The questions require you to carry out directions involving the information given in the chart.

6) Skills and aptitudes

Performance tests effectively measure some manual skills and aptitudes. When the skill is one in which you are trained, such as typing or shorthand, you can practice. These tests are often very much like those given in business school or high school courses. For many of the other skills and aptitudes, however, no short-time preparation can be made. Skills and abilities natural to you or that you have developed throughout your lifetime are being tested.

Many of the general questions just described provide all the data needed to answer the questions and ask you to use your reasoning ability to find the answers. Your best preparation for these tests, as well as for tests of facts and ideas, is to be at your physical and mental best. You, no doubt, have your own methods of getting into an exam-taking mood and keeping "in shape." The next section lists some ideas on this subject.

IV. KINDS OF QUESTIONS

Only rarely is the "essay" question, which you answer in narrative form, used in civil service tests. Civil service tests are usually of the short-answer type. Full instructions for answering these questions will be given to you at the examination. But in case this is your first experience with short-answer questions and separate answer sheets, here is what you need to know:

1) Multiple-choice Questions

Most popular of the short-answer questions is the "multiple choice" or "best answer" question. It can be used, for example, to test for factual knowledge, ability to solve problems or judgment in meeting situations found at work.

A multiple-choice question is normally one of three types—

- It can begin with an incomplete statement followed by several possible endings. You are to find the one ending which *best* completes the statement, although some of the others may not be entirely wrong.
- It can also be a complete statement in the form of a question which is answered by choosing one of the statements listed.

- It can be in the form of a problem – again you select the best answer.

Here is an example of a multiple-choice question with a discussion which should give you some clues as to the method for choosing the right answer:

When an employee has a complaint about his assignment, the action which will *best* help him overcome his difficulty is to

A. discuss his difficulty with his coworkers
B. take the problem to the head of the organization
C. take the problem to the person who gave him the assignment
D. say nothing to anyone about his complaint

In answering this question, you should study each of the choices to find which is best. Consider choice "A" – Certainly an employee may discuss his complaint with fellow employees, but no change or improvement can result, and the complaint remains unresolved. Choice "B" is a poor choice since the head of the organization probably does not know what assignment you have been given, and taking your problem to him is known as "going over the head" of the supervisor. The supervisor, or person who made the assignment, is the person who can clarify it or correct any injustice. Choice "C" is, therefore, correct. To say nothing, as in choice "D," is unwise. Supervisors have and interest in knowing the problems employees are facing, and the employee is seeking a solution to his problem.

2) True/False Questions

The "true/false" or "right/wrong" form of question is sometimes used. Here a complete statement is given. Your job is to decide whether the statement is right or wrong.

SAMPLE: A roaming cell-phone call to a nearby city costs less than a non-roaming call to a distant city.

This statement is wrong, or false, since roaming calls are more expensive.

This is not a complete list of all possible question forms, although most of the others are variations of these common types. You will always get complete directions for answering questions. Be sure you understand *how* to mark your answers – ask questions until you do.

V. RECORDING YOUR ANSWERS

Computer terminals are used more and more today for many different kinds of exams.

For an examination with very few applicants, you may be told to record your answers in the test booklet itself. Separate answer sheets are much more common. If this separate answer sheet is to be scored by machine – and this is often the case – it is highly important that you mark your answers correctly in order to get credit.

An electronic scoring machine is often used in civil service offices because of the speed with which papers can be scored. Machine-scored answer sheets must be marked with a pencil, which will be given to you. This pencil has a high graphite content which responds to the electronic scoring machine. As a matter of fact, stray dots may register as answers, so do not let your pencil rest on the answer sheet while you are pondering the correct answer. Also, if your pencil lead breaks or is otherwise defective, ask for another.

Since the answer sheet will be dropped in a slot in the scoring machine, be careful not to bend the corners or get the paper crumpled.

The answer sheet normally has five vertical columns of numbers, with 30 numbers to a column. These numbers correspond to the question numbers in your test booklet. After each number, going across the page are four or five pairs of dotted lines. These short dotted lines have small letters or numbers above them. The first two pairs may also have a "T" or "F" above the letters. This indicates that the first two pairs only are to be used if the questions are of the true-false type. If the questions are multiple choice, disregard the "T" and "F" and pay attention only to the small letters or numbers.

Answer your questions in the manner of the sample that follows:

32. The largest city in the United States is
    A. Washington, D.C.
    B. New York City
    C. Chicago
    D. Detroit
    E. San Francisco

1) Choose the answer you think is best. (New York City is the largest, so "B" is correct.)
2) Find the row of dotted lines numbered the same as the question you are answering. (Find row number 32)
3) Find the pair of dotted lines corresponding to the answer. (Find the pair of lines under the mark "B.")
4) Make a solid black mark between the dotted lines.

VI. BEFORE THE TEST

Common sense will help you find procedures to follow to get ready for an examination. Too many of us, however, overlook these sensible measures. Indeed, nervousness and fatigue have been found to be the most serious reasons why applicants fail to do their best on civil service tests. Here is a list of reminders:

- Begin your preparation early – Don't wait until the last minute to go scurrying around for books and materials or to find out what the position is all about.
- Prepare continuously – An hour a night for a week is better than an all-night cram session. This has been definitely established. What is more, a night a week for a month will return better dividends than crowding your study into a shorter period of time.
- Locate the place of the exam – You have been sent a notice telling you when and where to report for the examination. If the location is in a different town or otherwise unfamiliar to you, it would be well to inquire the best route and learn something about the building.
- Relax the night before the test – Allow your mind to rest. Do not study at all that night. Plan some mild recreation or diversion; then go to bed early and get a good night's sleep.
- Get up early enough to make a leisurely trip to the place for the test – This way unforeseen events, traffic snarls, unfamiliar buildings, etc. will not upset you.
- Dress comfortably – A written test is not a fashion show. You will be known by number and not by name, so wear something comfortable.

- Leave excess paraphernalia at home – Shopping bags and odd bundles will get in your way. You need bring only the items mentioned in the official notice you received; usually everything you need is provided. Do not bring reference books to the exam. They will only confuse those last minutes and be taken away from you when in the test room.
- Arrive somewhat ahead of time – If because of transportation schedules you must get there very early, bring a newspaper or magazine to take your mind off yourself while waiting.
- Locate the examination room – When you have found the proper room, you will be directed to the seat or part of the room where you will sit. Sometimes you are given a sheet of instructions to read while you are waiting. Do not fill out any forms until you are told to do so; just read them and be prepared.
- Relax and prepare to listen to the instructions
- If you have any physical problem that may keep you from doing your best, be sure to tell the test administrator. If you are sick or in poor health, you really cannot do your best on the exam. You can come back and take the test some other time.

VII. AT THE TEST

The day of the test is here and you have the test booklet in your hand. The temptation to get going is very strong. Caution! There is more to success than knowing the right answers. You must know how to identify your papers and understand variations in the type of short-answer question used in this particular examination. Follow these suggestions for maximum results from your efforts:

1) Cooperate with the monitor

The test administrator has a duty to create a situation in which you can be as much at ease as possible. He will give instructions, tell you when to begin, check to see that you are marking your answer sheet correctly, and so on. He is not there to guard you, although he will see that your competitors do not take unfair advantage. He wants to help you do your best.

2) Listen to all instructions

Don't jump the gun! Wait until you understand all directions. In most civil service tests you get more time than you need to answer the questions. So don't be in a hurry. Read each word of instructions until you clearly understand the meaning. Study the examples, listen to all announcements and follow directions. Ask questions if you do not understand what to do.

3) Identify your papers

Civil service exams are usually identified by number only. You will be assigned a number; you must not put your name on your test papers. Be sure to copy your number correctly. Since more than one exam may be given, copy your exact examination title.

4) Plan your time

Unless you are told that a test is a "speed" or "rate of work" test, speed itself is usually not important. Time enough to answer all the questions will be provided, but this does not mean that you have all day. An overall time limit has been set. Divide the total time (in minutes) by the number of questions to determine the approximate time you have for each question.

5) Do not linger over difficult questions

If you come across a difficult question, mark it with a paper clip (useful to have along) and come back to it when you have been through the booklet. One caution if you do this – be sure to skip a number on your answer sheet as well. Check often to be sure that you have not lost your place and that you are marking in the row numbered the same as the question you are answering.

6) Read the questions

Be sure you know what the question asks! Many capable people are unsuccessful because they failed to *read* the questions correctly.

7) Answer all questions

Unless you have been instructed that a penalty will be deducted for incorrect answers, it is better to guess than to omit a question.

8) Speed tests

It is often better NOT to guess on speed tests. It has been found that on timed tests people are tempted to spend the last few seconds before time is called in marking answers at random – without even reading them – in the hope of picking up a few extra points. To discourage this practice, the instructions may warn you that your score will be "corrected" for guessing. That is, a penalty will be applied. The incorrect answers will be deducted from the correct ones, or some other penalty formula will be used.

9) Review your answers

If you finish before time is called, go back to the questions you guessed or omitted to give them further thought. Review other answers if you have time.

10) Return your test materials

If you are ready to leave before others have finished or time is called, take ALL your materials to the monitor and leave quietly. Never take any test material with you. The monitor can discover whose papers are not complete, and taking a test booklet may be grounds for disqualification.

VIII. EXAMINATION TECHNIQUES

1) Read the general instructions carefully. These are usually printed on the first page of the exam booklet. As a rule, these instructions refer to the timing of the examination; the fact that you should not start work until the signal and must stop work at a signal, etc. If there are any *special* instructions, such as a choice of questions to be answered, make sure that you note this instruction carefully.

2) When you are ready to start work on the examination, that is as soon as the signal has been given, read the instructions to each question booklet, underline any key words or phrases, such as *least*, *best*, *outline*, *describe* and the like. In this way you will tend to answer as requested rather than discover on reviewing your paper that you *listed without describing*, that you selected the *worst* choice rather than the *best* choice, etc.

3) If the examination is of the objective or multiple-choice type – that is, each question will also give a series of possible answers: A, B, C or D, and you are called upon to select the best answer and write the letter next to that answer on your answer paper – it is advisable to start answering each question in turn. There may be anywhere from 50 to 100 such questions in the three or four hours allotted and you can see how much time would be taken if you read through all the questions before beginning to answer any. Furthermore, if you come across a question or group of questions which you know would be difficult to answer, it would undoubtedly affect your handling of all the other questions.

4) If the examination is of the essay type and contains but a few questions, it is a moot point as to whether you should read all the questions before starting to answer any one. Of course, if you are given a choice – say five out of seven and the like – then it is essential to read all the questions so you can eliminate the two that are most difficult. If, however, you are asked to answer all the questions, there may be danger in trying to answer the easiest one first because you may find that you will spend too much time on it. The best technique is to answer the first question, then proceed to the second, etc.

5) Time your answers. Before the exam begins, write down the time it started, then add the time allowed for the examination and write down the time it must be completed, then divide the time available somewhat as follows:
   - If 3-1/2 hours are allowed, that would be 210 minutes. If you have 80 objective-type questions, that would be an average of 2-1/2 minutes per question. Allow yourself no more than 2 minutes per question, or a total of 160 minutes, which will permit about 50 minutes to review.
   - If for the time allotment of 210 minutes there are 7 essay questions to answer, that would average about 30 minutes a question. Give yourself only 25 minutes per question so that you have about 35 minutes to review.

6) The most important instruction is to *read each question* and make sure you know what is wanted. The second most important instruction is to *time yourself properly* so that you answer every question. The third most important instruction is to *answer every question*. Guess if you have to but include something for each question. Remember that you will receive no credit for a blank and will probably receive some credit if you write something in answer to an essay question. If you guess a letter – say "B" for a multiple-choice question – you may have guessed right. If you leave a blank as an answer to a multiple-choice question, the examiners may respect your feelings but it will not add a point to your score. Some exams may penalize you for wrong answers, so in such cases *only*, you may not want to guess unless you have some basis for your answer.

7) Suggestions
   a. Objective-type questions
      1. Examine the question booklet for proper sequence of pages and questions
      2. Read all instructions carefully
      3. Skip any question which seems too difficult; return to it after all other questions have been answered
      4. Apportion your time properly; do not spend too much time on any single question or group of questions

5. Note and underline key words – *all, most, fewest, least, best, worst, same, opposite,* etc.
6. Pay particular attention to negatives
7. Note unusual option, e.g., unduly long, short, complex, different or similar in content to the body of the question
8. Observe the use of "hedging" words – *probably, may, most likely,* etc.
9. Make sure that your answer is put next to the same number as the question
10. Do not second-guess unless you have good reason to believe the second answer is definitely more correct
11. Cross out original answer if you decide another answer is more accurate; do not erase until you are ready to hand your paper in
12. Answer all questions; guess unless instructed otherwise
13. Leave time for review

b. Essay questions

1. Read each question carefully
2. Determine exactly what is wanted. Underline key words or phrases.
3. Decide on outline or paragraph answer
4. Include many different points and elements unless asked to develop any one or two points or elements
5. Show impartiality by giving pros and cons unless directed to select one side only
6. Make and write down any assumptions you find necessary to answer the questions
7. Watch your English, grammar, punctuation and choice of words
8. Time your answers; don't crowd material

8) Answering the essay question

Most essay questions can be answered by framing the specific response around several key words or ideas. Here are a few such key words or ideas:

M's: manpower, materials, methods, money, management
P's: purpose, program, policy, plan, procedure, practice, problems, pitfalls, personnel, public relations

a. Six basic steps in handling problems:

1. Preliminary plan and background development
2. Collect information, data and facts
3. Analyze and interpret information, data and facts
4. Analyze and develop solutions as well as make recommendations
5. Prepare report and sell recommendations
6. Install recommendations and follow up effectiveness

b. Pitfalls to avoid

1. *Taking things for granted* – A statement of the situation does not necessarily imply that each of the elements is necessarily true; for example, a complaint may be invalid and biased so that all that can be taken for granted is that a complaint has been registered

2. *Considering only one side of a situation* – Wherever possible, indicate several alternatives and then point out the reasons you selected the best one
3. *Failing to indicate follow up* – Whenever your answer indicates action on your part, make certain that you will take proper follow-up action to see how successful your recommendations, procedures or actions turn out to be
4. *Taking too long in answering any single question* – Remember to time your answers properly

IX. AFTER THE TEST

Scoring procedures differ in detail among civil service jurisdictions although the general principles are the same. Whether the papers are hand-scored or graded by machine we have described, they are nearly always graded by number. That is, the person who marks the paper knows only the number – never the name – of the applicant. Not until all the papers have been graded will they be matched with names. If other tests, such as training and experience or oral interview ratings have been given, scores will be combined. Different parts of the examination usually have different weights. For example, the written test might count 60 percent of the final grade, and a rating of training and experience 40 percent. In many jurisdictions, veterans will have a certain number of points added to their grades.

After the final grade has been determined, the names are placed in grade order and an eligible list is established. There are various methods for resolving ties between those who get the same final grade – probably the most common is to place first the name of the person whose application was received first. Job offers are made from the eligible list in the order the names appear on it. You will be notified of your grade and your rank as soon as all these computations have been made. This will be done as rapidly as possible.

People who are found to meet the requirements in the announcement are called "eligibles." Their names are put on a list of eligible candidates. An eligible's chances of getting a job depend on how high he stands on this list and how fast agencies are filling jobs from the list.

When a job is to be filled from a list of eligibles, the agency asks for the names of people on the list of eligibles for that job. When the civil service commission receives this request, it sends to the agency the names of the three people highest on this list. Or, if the job to be filled has specialized requirements, the office sends the agency the names of the top three persons who meet these requirements from the general list.

The appointing officer makes a choice from among the three people whose names were sent to him. If the selected person accepts the appointment, the names of the others are put back on the list to be considered for future openings.

That is the rule in hiring from all kinds of eligible lists, whether they are for typist, carpenter, chemist, or something else. For every vacancy, the appointing officer has his choice of any one of the top three eligibles on the list. This explains why the person whose name is on top of the list sometimes does not get an appointment when some of the persons lower on the list do. If the appointing officer chooses the second or third eligible, the No. 1 eligible does not get a job at once, but stays on the list until he is appointed or the list is terminated.

X. HOW TO PASS THE INTERVIEW TEST

The examination for which you applied requires an oral interview test. You have already taken the written test and you are now being called for the interview test – the final part of the formal examination.

You may think that it is not possible to prepare for an interview test and that there are no procedures to follow during an interview. Our purpose is to point out some things you can do in advance that will help you and some good rules to follow and pitfalls to avoid while you are being interviewed.

What is an interview supposed to test?

The written examination is designed to test the technical knowledge and competence of the candidate; the oral is designed to evaluate intangible qualities, not readily measured otherwise, and to establish a list showing the relative fitness of each candidate – as measured against his competitors – for the position sought. Scoring is not on the basis of "right" and "wrong," but on a sliding scale of values ranging from "not passable" to "outstanding." As a matter of fact, it is possible to achieve a relatively low score without a single "incorrect" answer because of evident weakness in the qualities being measured.

Occasionally, an examination may consist entirely of an oral test – either an individual or a group oral. In such cases, information is sought concerning the technical knowledges and abilities of the candidate, since there has been no written examination for this purpose. More commonly, however, an oral test is used to supplement a written examination.

Who conducts interviews?

The composition of oral boards varies among different jurisdictions. In nearly all, a representative of the personnel department serves as chairman. One of the members of the board may be a representative of the department in which the candidate would work. In some cases, "outside experts" are used, and, frequently, a businessman or some other representative of the general public is asked to serve. Labor and management or other special groups may be represented. The aim is to secure the services of experts in the appropriate field.

However the board is composed, it is a good idea (and not at all improper or unethical) to ascertain in advance of the interview who the members are and what groups they represent. When you are introduced to them, you will have some idea of their backgrounds and interests, and at least you will not stutter and stammer over their names.

What should be done before the interview?

While knowledge about the board members is useful and takes some of the surprise element out of the interview, there is other preparation which is more substantive. It *is* possible to prepare for an oral interview – in several ways:

1) Keep a copy of your application and review it carefully before the interview

This may be the only document before the oral board, and the starting point of the interview. Know what education and experience you have listed there, and the sequence and dates of all of it. Sometimes the board will ask you to review the highlights of your experience for them; you should not have to hem and haw doing it.

2) Study the class specification and the examination announcement

Usually, the oral board has one or both of these to guide them. The qualities, characteristics or knowledges required by the position sought are stated in these documents. They offer valuable clues as to the nature of the oral interview. For example, if the job involves supervisory responsibilities, the announcement will usually indicate that knowledge of modern supervisory methods and the qualifications of the candidate as a supervisor will be tested. If so, you can expect such questions, frequently in the form of a hypothetical situation which you are expected to solve. NEVER go into an oral without knowledge of the duties and responsibilities of the job you seek.

3) Think through each qualification required

Try to visualize the kind of questions you would ask if you were a board member. How well could you answer them? Try especially to appraise your own knowledge and background in each area, *measured against the job sought*, and identify any areas in which you are weak. Be critical and realistic – do not flatter yourself.

4) Do some general reading in areas in which you feel you may be weak

For example, if the job involves supervision and your past experience has NOT, some general reading in supervisory methods and practices, particularly in the field of human relations, might be useful. Do NOT study agency procedures or detailed manuals. The oral board will be testing your understanding and capacity, not your memory.

5) Get a good night's sleep and watch your general health and mental attitude

You will want a clear head at the interview. Take care of a cold or any other minor ailment, and of course, no hangovers.

What should be done on the day of the interview?

Now comes the day of the interview itself. Give yourself plenty of time to get there. Plan to arrive somewhat ahead of the scheduled time, particularly if your appointment is in the fore part of the day. If a previous candidate fails to appear, the board might be ready for you a bit early. By early afternoon an oral board is almost invariably behind schedule if there are many candidates, and you may have to wait. Take along a book or magazine to read, or your application to review, but leave any extraneous material in the waiting room when you go in for your interview. In any event, relax and compose yourself.

The matter of dress is important. The board is forming impressions about you – from your experience, your manners, your attitude, and your appearance. Give your personal appearance careful attention. Dress your best, but not your flashiest. Choose conservative, appropriate clothing, and be sure it is immaculate. This is a business interview, and your appearance should indicate that you regard it as such. Besides, being well groomed and properly dressed will help boost your confidence.

Sooner or later, someone will call your name and escort you into the interview room. *This is it.* From here on you are on your own. It is too late for any more preparation. But remember, you asked for this opportunity to prove your fitness, and you are here because your request was granted.

What happens when you go in?

The usual sequence of events will be as follows: The clerk (who is often the board stenographer) will introduce you to the chairman of the oral board, who will introduce you to the other members of the board. Acknowledge the introductions before you sit down. Do not be surprised if you find a microphone facing you or a stenotypist sitting by. Oral interviews are usually recorded in the event of an appeal or other review.

Usually the chairman of the board will open the interview by reviewing the highlights of your education and work experience from your application – primarily for the benefit of the other members of the board, as well as to get the material into the record. Do not interrupt or comment unless there is an error or significant misinterpretation; if that is the case, do not hesitate. But do not quibble about insignificant matters. Also, he will usually ask you some question about your education, experience or your present job – partly to get you to start talking and to establish the interviewing "rapport." He may start the actual questioning, or turn it over to one of the other members. Frequently, each member undertakes the questioning on a particular area, one in which he is perhaps most competent, so you can expect each member to participate in the examination. Because time is limited, you may also expect some rather abrupt switches in the direction the questioning takes, so do not be upset by it. Normally, a board member will not pursue a single line of questioning unless he discovers a particular strength or weakness.

After each member has participated, the chairman will usually ask whether any member has any further questions, then will ask you if you have anything you wish to add. Unless you are expecting this question, it may floor you. Worse, it may start you off on an extended, extemporaneous speech. The board is not usually seeking more information. The question is principally to offer you a last opportunity to present further qualifications or to indicate that you have nothing to add. So, if you feel that a significant qualification or characteristic has been overlooked, it is proper to point it out in a sentence or so. Do not compliment the board on the thoroughness of their examination – they have been sketchy, and you know it. If you wish, merely say, "No thank you, I have nothing further to add." This is a point where you can "talk yourself out" of a good impression or fail to present an important bit of information. Remember, *you close the interview yourself.*

The chairman will then say, "That is all, Mr. _______, thank you." Do not be startled; the interview is over, and quicker than you think. Thank him, gather your belongings and take your leave. Save your sigh of relief for the other side of the door.

How to put your best foot forward

Throughout this entire process, you may feel that the board individually and collectively is trying to pierce your defenses, seek out your hidden weaknesses and embarrass and confuse you. Actually, this is not true. They are obliged to make an appraisal of your qualifications for the job you are seeking, and they want to see you in your best light. Remember, they must interview all candidates and a non-cooperative candidate may become a failure in spite of their best efforts to bring out his qualifications. Here are 15 suggestions that will help you:

1) Be natural – Keep your attitude confident, not cocky

If you are not confident that you can do the job, do not expect the board to be. Do not apologize for your weaknesses, try to bring out your strong points. The board is interested in a positive, not negative, presentation. Cockiness will antagonize any board member and make him wonder if you are covering up a weakness by a false show of strength.

2) Get comfortable, but don't lounge or sprawl

Sit erectly but not stiffly. A careless posture may lead the board to conclude that you are careless in other things, or at least that you are not impressed by the importance of the occasion. Either conclusion is natural, even if incorrect. Do not fuss with your clothing, a pencil or an ashtray. Your hands may occasionally be useful to emphasize a point; do not let them become a point of distraction.

3) Do not wisecrack or make small talk

This is a serious situation, and your attitude should show that you consider it as such. Further, the time of the board is limited – they do not want to waste it, and neither should you.

4) Do not exaggerate your experience or abilities

In the first place, from information in the application or other interviews and sources, the board may know more about you than you think. Secondly, you probably will not get away with it. An experienced board is rather adept at spotting such a situation, so do not take the chance.

5) If you know a board member, do not make a point of it, yet do not hide it

Certainly you are not fooling him, and probably not the other members of the board. Do not try to take advantage of your acquaintanceship – it will probably do you little good.

6) Do not dominate the interview

Let the board do that. They will give you the clues – do not assume that you have to do all the talking. Realize that the board has a number of questions to ask you, and do not try to take up all the interview time by showing off your extensive knowledge of the answer to the first one.

7) Be attentive

You only have 20 minutes or so, and you should keep your attention at its sharpest throughout. When a member is addressing a problem or question to you, give him your undivided attention. Address your reply principally to him, but do not exclude the other board members.

8) Do not interrupt

A board member may be stating a problem for you to analyze. He will ask you a question when the time comes. Let him state the problem, and wait for the question.

9) Make sure you understand the question

Do not try to answer until you are sure what the question is. If it is not clear, restate it in your own words or ask the board member to clarify it for you. However, do not haggle about minor elements.

10) Reply promptly but not hastily

A common entry on oral board rating sheets is "candidate responded readily," or "candidate hesitated in replies." Respond as promptly and quickly as you can, but do not jump to a hasty, ill-considered answer.

11) Do not be peremptory in your answers

A brief answer is proper – but do not fire your answer back. That is a losing game from your point of view. The board member can probably ask questions much faster than you can answer them.

12) Do not try to create the answer you think the board member wants

He is interested in what kind of mind you have and how it works – not in playing games. Furthermore, he can usually spot this practice and will actually grade you down on it.

13) Do not switch sides in your reply merely to agree with a board member

Frequently, a member will take a contrary position merely to draw you out and to see if you are willing and able to defend your point of view. Do not start a debate, yet do not surrender a good position. If a position is worth taking, it is worth defending.

14) Do not be afraid to admit an error in judgment if you are shown to be wrong

The board knows that you are forced to reply without any opportunity for careful consideration. Your answer may be demonstrably wrong. If so, admit it and get on with the interview.

15) Do not dwell at length on your present job

The opening question may relate to your present assignment. Answer the question but do not go into an extended discussion. You are being examined for a *new* job, not your present one. As a matter of fact, try to phrase ALL your answers in terms of the job for which you are being examined.

Basis of Rating

Probably you will forget most of these "do's" and "don'ts" when you walk into the oral interview room. Even remembering them all will not ensure you a passing grade. Perhaps you did not have the qualifications in the first place. But remembering them will help you to put your best foot forward, without treading on the toes of the board members.

Rumor and popular opinion to the contrary notwithstanding, an oral board wants you to make the best appearance possible. They know you are under pressure – but they also want to see how you respond to it as a guide to what your reaction would be under the pressures of the job you seek. They will be influenced by the degree of poise you display, the personal traits you show and the manner in which you respond.

ABOUT THIS BOOK

This book contains tests divided into Examination Sections. Go through each test, answering every question in the margin. We have also attached a sample answer sheet at the back of the book that can be removed and used. At the end of each test look at the answer key and check your answers. On the ones you got wrong, look at the right answer choice and learn. Do not fill in the answers first. Do not memorize the questions and answers, but understand the answer and principles involved. On your test, the questions will likely be different from the samples. Questions are changed and new ones added. If you understand these past questions you should have success with any changes that arise. Tests may consist of several types of questions. We have additional books on each subject should more study be advisable or necessary for you. Finally, the more you study, the better prepared you will be. This book is intended to be the last thing you study before you walk into the examination room. Prior study of relevant texts is also recommended. NLC publishes some of these in our Fundamental Series. Knowledge and good sense are important factors in passing your exam. Good luck also helps. So now study this Passbook, absorb the material contained within and take that knowledge into the examination. Then do your best to pass that exam.

EXAMINATION SECTION

EXAMINATION SECTION
TEST 1

DIRECTIONS: Each question or incomplete statement is followed by several suggested answers or completions. Select the one that BEST answers the question or completes the statement. *PRINT THE LETTER OF THE CORRECT ANSWER IN THE SPACE AT THE RIGHT.*

1. The rate at which solids settle out of sewage in a sedimentation tank is dependent MAINLY on the 1.____

   A. depth of sewage in tank
   B. velocity of flow through tank
   C. water pressure
   D. amount of solids in sewage

2. Flow of sewage to the treatment plant from the intercepting sewer is controlled by a 2.____

   A. sluice gate
   B. flight
   C. reduction valve
   D. bar screen

3. The process of adding chemicals to the sewage to increase the rate of settlement of suspended solids is known as 3.____

   A. calcination
   B. oxydation
   C. flocculation
   D. chlorination

4. Large objects, such as sticks, are removed from raw sewage by a 4.____

   A. sludge pump
   B. settling tank
   C. bar rack
   D. ejector

5. Grit is MOST frequently moved from the grit chamber to the grit storage tank by 5.____

   A. gravity flow
   B. compressed air
   C. wheelbarrow
   D. conveyor belt

6. The porous plates through which air enters the aeration chamber in the activates sludge process are known as 6.____

   A. diffusers
   B. nozzles
   C. oxygen lances
   D. pressure plates

7. The *strength* of sewage is measured by determining its 7.____

   A. M.D.
   B. HP
   C. G.P.M.
   D. B.O.D.

8. In order to prevent digestion of sludge in sedimentation tanks, the sludge is 8.____

   A. chemically treated
   B. aerated
   C. continuously removed
   D. heated

9. One of the chemicals used to increase the rate of settling of suspended solids in sewage is 9.____

   A. bromine
   B. carbon
   C. copper
   D. fluorine

10. Chlorine is added to sewage to 10.____

A. aid sludge digestion
B. kill bacteria
C. increase B.T.U. content of gas
D. dewater sludge in storage tanks

11. A sewer which receives BOTH rain water and sewage from residences is known as a _______ sewer. 11.____

A. storm B. sanitary C. regulated D. combined

12. Treated sewage flowing out of the sewage treatment plant is known as the 12.____

A. desiccant B. decanter C. effluent D. waste

13. In the aerated sludge process, grease can conveniently be removed from the sewage in the 13.____

A. wet well
B. final sedimentation chamber
C. grit chamber
D. screening chamber

14. A device used to control the rate of flow of sewage is known as a 14.____

A. weir B. penstock C. agitator D. stator

15. A venturi meter is used to measure _______ of sewage. 15.____

A. pressure B. temperature
C. flow D. depth

16. Sludge gas is composed MAINLY of 16.____

A. methane B. carbon monoxide
C. hydrogen sulfide D. ammonia

17. Overloads on reciprocating pumps can be prevented by _______ valves. 17.____

A. check B. relief C. gate D. globe

18. A valve that permits flow in only one direction is a _______ valve. 18.____

A. check B. plug C. gate D. globe

19. Where a quick closing action is desired, the type of valve that should be used is a(n) 19.____

A. globe B. needle C. gate D. angle

20. The one of the following types of valves that causes the LEAST resistance to the flow of sewage is a(n) 20.____

A. globe B. angle C. gate D. key

21. Cavitation (pitting) in centrifugal pumps would MOST probably occur in the 21.____

A. packing glands B. roller bearings
C. pump impellers D. shaft

22. The type of pump MOST commonly used to pump sludge is the 22.____

A. turbine
B. centrifugal
C. volute
D. reciprocating

23. If the bearings on a large pump become excessively hot, the BEST thing to do is to 23.____

A. pour cold water on the bearings to cool them
B. fill the oil cup and slow the pump till the bearings cool
C. stop the motor and check the condition of the bearings and oil or grease
D. shunt the suction and discharge valves till the pump cools, then reopen valves slowly

24. Priming is MOST frequently required in a ______ pump. 24.____

A. centrifugal
B. reciprocating
C. rotary
D. gear

25. When gland nuts on a sewage pump are properly tightened, 25.____

A. there will be slight leakage through the packing
B. the packing is not compressed
C. the lantern will rotate
D. the stuffing box cannot overheat

KEY (CORRECT ANSWERS)

1.	B	11.	D
2.	A	12.	C
3.	C	13.	B
4.	C	14.	A
5.	B	15.	C
6.	A	16.	A
7.	D	17.	B
8.	C	18.	A
9.	C	19.	C
10.	B	20.	C

21. C
22. D
23. C
24. A
25. A

TEST 2

DIRECTIONS: Each question or incomplete statement is followed by several suggested answers or completions. Select the one that BEST answers the question or completes the statement. *PRINT THE LETTER OF THE CORRECT ANSWER IN THE SPACE AT THE RIGHT.*

1. The rated capacity of a pump is usually given in terms of 1.____

   A. horsepower and velocity of flow
   B. gallons per minute pumped and pressure head
   C. electrical consumption and velocity of flow
   D. electrical consumption and gallons per minute pumped

2. Sudden shutting of the discharge valve of a centrifugal pump may damage a piping system because 2.____

   A. the impellers will turn for a short period without lubrication
   B. there will be leakage past the packing
   C. pump bearings will be scored
   D. water hammer will occur

3. The one of the following that is used on a piece of mechanical equipment to prevent overloading is a 3.____

   A. bushing
   B. shear pin
   C. split ring
   D. yoke

4. Sewage has suddenly stopped flowing from a centrifugal pump which has been working well. 4.____
   The MOST probable cause for this is that

   A. air is leaking into suction line
   B. the bearings are worn
   C. the speed of pump is excessive
   D. the oil level is inadequate

5. The MOST viscous of the following lubricants is 5.____

   A. diesel oil
   B. cup grease
   C. S.A.E. 40 oil
   D. kerosene

6. The BEST method of lubricating roller bearings if by means of 6.____

   A. light machine oil
   B. instrument oil
   C. diesel oil
   D. grease

7. A pipe reducer would be used to 7.____

   A. permit drawing of low pressure gas from high pressure pipes
   B. connect two lines of different sizes
   C. compress packing in a line expansion joint
   D. remove excess water from sludge lines

8. The one of the following that is NOT a standard pipe fitting is a 8.____

A. union B. tap C. tee D. cross

9. Water hammer can be reduced by using a(n) 9.____

A. quick closing valve B. air chamber
C. flanged connection D. automatic primer

10. To be watertight, the faces of a flanged connection should be 10.____

A. machined
B. packed with waste
C. coated with rubber cement
D. etched

11. When tightening bolts on a flanged connection, the PROPER procedure is to 11.____

A. take up each bolt wrench hard before beginning to tighten next adjacent bolt
B. first take up one bolt wrench hard, then tighten diagonally opposite bolt wrench hard
C. take up each bolt gradually, tightening adjacent bolts in order
D. take up each bolt gradually, first tightening one bolt slightly, then the diagonally opposite bolt in a like manner

12. The kind of wrench used to tighten pipe would be a(n) 12.____

A. crescent B. open end C. monkey D. Stillson

13. The one of the following fire extinguisher types that should be used on a fire in an electric motor is 13.____

A. carbon dioxide B. soda acid
C. water fog D. foam

14. A safety device used to protect electrical circuits from overloads is a(n) 14.____

A. solenoid B. powerstat
C. circuit breaker D. transformer

15. The one of the following that is the MOST common reason for noisy operation of electric motors is 15.____

A. shorted windings B. worn brushes
C. worn bearings D. overloading

16. The unit of measurement used for determining the amount of energy consumed in running a motor is 16.____

A. volt-amperes B. kilowatt hours
C. horsepower D. frequency-cycle

17. The one of the following parts of an electric motor that will wear out MOST frequently is(are) the 17.____

A. armature B. field pieces
C. shaft D. brushes

18. The MOST important reason for using a fuse in an electrical circuit is to prevent excessive 18.____

A. voltage
B. current
C. frequency
D. resistance

19. An electrical device used to increase line voltage is a(n) 19.____

A. alternator
B. magneto
C. transformer
D. choke

20. The minimum size wire that should be used to supply power to a 1 H.P. motor is 20.____

A. #14 B. #16 C. #18 D. #20

21. The one of the following that should be used to clean a commit at or is 21.____

A. sandpaper
B. emery paper
C. pumice
D. emery cloth

22. To clean armature windings on a motor, one should use 22.____

A. calcium chloride
B. sodium chloride
C. carbon tetrachloride
D. sodium hypochlorite

23. When starting an electrically driven centrifugal pump, the starting load can be reduced by 23.____

A. opening the suction and discharge valves
B. opening the suction valve and closing the discharge valves
C. closing the suction and discharge valves
D. closing the suction valve and opening the discharge valves

24. For increased safety, the frame of an electric motor should be 24.____

A. grounded B. shorted C. shunted D. painted

25. The one of the following materials that is MOST suitable for piping corrosive sludge gas is 25.____

A. copper B. steel C. aluminum D. transite

KEY (CORRECT ANSWERS)

1. B
2. D
3. B
4. A
5. B
6. D
7. B
8. B
9. B
10. A
11. D
12. D
13. A
14. C
15. C
16. B
17. D
18. B
19. C
20. A
21. A
22. C
23. B
24. A
25. D

EXAMINATION SECTION
TEST 1

DIRECTIONS: Answer the following questions directly, briefly, and succinctly.

Questions 1-10.

DIRECTIONS: What is the purpose of each of the following pieces of equipment in sewage treatment?

1. Coarse racks
2. Fine bar screens
3. Fine screens
4. Screening grinders
5. Grit chambers
6. Grit washers
7. Settling tanks
8. Aeration tanks
9. Sludge digesters
10. Sluice gates or weirs

Questions 11-20.

DIRECTIONS: What are the chief causes for each of the following pieces of equipment becoming defective? For each cause, indicate how you would repair the defect.

11. Mechanical screens
12. Screening grinders
13. Grit collectors
14. Grit washers
15. Main sewage and circulating pumps
16. Sludge pumps
17. Settling tanks
18. Aeration tanks
19. Sluice gates
20. Storage tanks

Questions 21-25.

21. What safety precaution should you take when making repairs on electric equipment?

22. What safety precaution should you take when making repairs on a mechanical rack?

23. What safety precaution should you take when grinding tools?

24. What safety precaution should you take when working with digester gas?

25. What safety precaution should you take when handling a leaking chlorine cylinder?

Questions 26-30.

26. How is grit removed from grit channels?

27. What should be done to protect flocculator mechanism in case of accumulation of grit and sand in the flocculators?

28. What is the usual period of detention in settling tanks?

29. Indicate, on a simple sketch, the direction of flow in settling tanks in use in the city.

30. What must be done in order to prevent digestion of sludge in settling tanks?

Questions 31-35.

31. In chemical precipitation, in what form are chemicals received for mixing with the sewage to increase settling?

32. In the activated sludge process, what is mixed with the raw or clarified sewage?

33. Where, in the activated sludge process, are diffuser blocks used?

34. Why is chlorine added to effluent discharging from some sewage treatment plants into bodies of water?

35. How are sludge digestion tanks heated?

Questions 36-40.

36. What is the average period of sludge detention in primary tanks?

37. What happens to sludge volume in secondary tanks?

38. What is usually done with the digested sludge resulting from plant operation?

39. What use is made of skimmings from settling tanks?

40. At what point in the activated sludge process is the sample for checking residual oxygen taken?

Questions 41-45.

41. For what is a venturi motor used?

42. What type of pump is generally used for pumping raw sewage?

43. What type of pump is generally used for pumping sludge?

44. How should electric motors be cleaned?

45. What type of fire extinguisher should be used for fires on electrical equipment?

Questions 46-50.

46. Assume that you are pumping water into a tank at the rate of 200 gallons per minute and that you are withdrawing water at the rate of 75 gallons per minute. How long will it take to add 60,000 gallons of water to the tank?

47.

The figures above show two readings of a watt-hour meter taken twenty-four hours apart. How many k.w. hours were used in that period?

KEY (CORRECT ANSWERS)

1. Remove coarse materials – cans, ashes, rags, timber, etc.
2. Remove smaller materials that go through coarse racks
3. Remove particles of sewage and fine floating material (screens also protect equipment of plant through which sewage must pass)
4. Reduce size of removed particles
5. Catch sand, gravel, ashes, and other gritty material
6. Remove odor-producing materials from the grit
7. Settle out materials in sewage (sludge)
8. Treat sewage by application of air
9. Digest sludge and produce gas
10. Control and measure flow of sewage
11. Overloading causes breakage of cables, chains, shear pins. Repair by removing cause of overload and replacing broken part.
12. Overloading through delivery of too much material or dulling of knives. Repair by opening grinder, removing obstructing material, and replacing or reversing knives.
13. Moving parts – chains, shoes, pins, sprockets, flights–becoming worn. Repair by replacing worn parts.
14. Clogging of slide valves. Repair by cleaning and replacing pipe fittings. Clean, repair, and adjust such parts as chains and diaphragms.
15. Motor failure. Report to plant engineer. Loss of prime and leakage. Adjust or replace packing, adjust water seal, and reprime.
16. Obstruction under valves, over-load, slogging or leaks. Repair by replacing faulty gaskets and worn packing for air leaks, to remove obstructions, take off valve covers, and clean.
17. Same as 13
18. Plates or valves becoming clogged or broken. Service valves, clean and replace clogged or broken plates.
19. Gate skewed or jammed, spindle bent, gears jammed or broken. Re-adjust or replace.
20. Same as 14.
21. Make sure power is off and equipment grounded
22. Make sure equipment cannot be set in motion
23. Wear goggles

24. Use proper mask, keep flame away

25. Wear special gas mask and rubber gloves

26. Longitudinal or revolving scrapers

27. Periodic flushing

28. One to two hours

29. 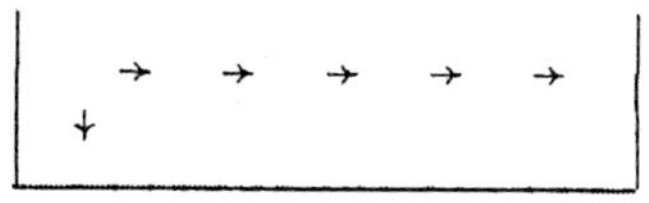

30. Continuous removal of settled sludge

31. Powder, dry with chlorine

32. Biologically active sludge, air (1/2)

33. Base of aeration tanks

34. Reduces harmful bacteria, minimizes disagreeable odors

35. Heat provided by circulating warm water through coils in the tanks

36. 20 days approximately for primary digestion tanks or 2 hours for primary settling tanks

37. Compacted to about 50% of original volume

38. Dumped at sea or used as landfill

39. Sold as fat

40. Between aeration tank and secondary settlement tank or at outlet of secondary settling tank

41. Measure flow of water

42. Centrifugal pump

43. Piston or reciprocating pump

44. Carbon tetrachloride, compressed air, rags (not waste)

45. CO_2, fire foam, carbon tetrachloride

46. 8 hours

47. 6583 − 4128 = 2455 or 6583 − 4129 = 2454

6583	or	6583
4128		4129
2455	or	2454

EXAMINATION SECTION

TEST 1

DIRECTIONS: Each question or incomplete statement is followed by several suggested answers or completions. Select the one that BEST answers the question or completes the statement. *PRINT THE LETTER OF THE CORRECT ANSWER IN THE SPACE AT THE RIGHT.*

1. Assume that a certain file has a safe edge. This is an edge that has 1.____

   A. no teeth
   B. the teeth pointing backward
   C. the teeth pointing forward
   D. fine criss-cross teeth

2. The one of the following which is the proper tool for threading a round rod is a 2.____

   A. tap B. countersink C. counterbore D. die

3. A rasp is a 3.____

   A. type of chisel
   B. type of file cleaner
   C. type of coarse file
   D. kind of plane

4. The one of the following which is the proper tool to use to tighten a round nut which has a series of notches cut in its outer surface is a(n) _______ wrench. 4.____

   A. box B. spanner C. Stillson D. monkey

5. Small leaks resulting from poor threads on steel or wrought-iron water pipes will often stop because the leaky threads are in time filled with 5.____

   A. sediment B. stalactite C. rust D. soapstone

6. A metal that can be rolled or beaten into very fine sheets is said to be 6.____

   A. anodized B. malleable C. tempered D. ferrous

7. To *rod* a sewer pipe means MOST likely to 7.____

   A. clean it out by means of rods
   B. keep the pipe clear of debris by placing a grating of rods at the intake
   C. support the sewer pipe with horizontal reinforcing metal rods
   D. shore up the pipe

8. Of the following abrasives, the one which is the LEAST coarse is 8.____

   A. No. 2 emery cloth
   B. crocus cloth
   C. No. 1 sandpaper
   D. No. 1/0 sandpaper

9. In order to permit free passage of water in one direction only and prevent a reversal of flow in the pipe, it is necessary to use a _______ valve. 9.____

   A. gate B. check C. globe D. needle

10. Assume that a cubic foot of water contains 7 1/2 gallons. The number of gallons of water which could be contained in a rectangular tank 3 feet long, 2 feet wide, and 2 feet deep is MOST NEARLY 10.____

    A. 12 B. 45 C. 90 D. 120

11. The weight, in pounds, of a cubic foot of fresh water is MOST NEARLY 11.____

A. 8.5 B. 32.4 C. 62.4 D. 98.6

12. The total weight, in pounds, of ten bags of Portland cement is MOST NEARLY _______ pounds. 12.____

A. 108 B. 187 C. 940 D. 1,200

13. If a concrete mix is said to be 1:2:4, this would mean that the mix is made up of 1 part by 1 13.____

A. volume of cement to 2 parts by volume of sand to 4 parts by volume of coarse aggregate
B. volume of cement to 2 parts by volume of coarse aggregate to 4 parts by volume of sand
C. volume of coarse aggregate to 2 parts by volume of sand to 4 parts by volume of cement
D. weight of cement to 2 parts by weight of coarse aggregate to 4 parts by weight of sand

14. The ratio of the weight of a substance to the weight of an equal volume of water is called the _______ of the substance. 14.____

A. specific volume
B. specific gravity
C. viscosity
D. fractional weight

15. Of the following, the pipe fitting which has four openings which permits connecting a line at right angles to another line is called a(n) 15.____

A. side outlet street L
B. double elbow
C. tee
D. cross

16. To tighten a nut where only a short swing of the wrench handle is possible, it is BEST to use a _______ wrench. 16.____

A. ratchet
B. hook spanner
C. Stillson
D. Bristo

17. Of the following, the proper tool to use to remove the burr from the inside of a pipe is a 17.____

A. half round file
B. reamer
C. mandrel
D. chisel

18. Fittings commonly used with copper pipe should be made of 18.____

A. brass
B. cast iron
C. malleable iron
D. pure tin

19. With respect to pipe, the abbreviation I.P.S. means 19.____

A. Internal Pipe Size
B. Iron Pipe Size
C. Iron Pipe Shape
D. International Pipe Size

20. A Stillson wrench is the proper wrench to use when tightening a 20.____

A. square nut
B. hexagonal nut
C. valve gland nut
D. pipe fitting

21. The one of the following which is the proper tool to use for cutting wood along the grain is a _______ saw. 21.____

A. rip
B. panel
C. cross-cut
D. back

22. The one of the following which is the proper tool to use to cut internal screw threads is a 22.____

A. broach
B. die
C. tap
D. stock

23. A center punch is the proper tool used to 23.____

A. cut out the center of a gasket
B. dent metal prior to drilling
C. drive nails beneath the surface of the wood
D. punch a small hole in sheet metal

24. The one of the following knots which can be safely used for tying together the ends of two dry ropes of the same size is a 24.____

A. granny knot
B. clove hitch
C. half hitch
D. square knot

25. The PRIMARY purpose of a trap under a plumbing fixture is to 25.____

A. act as a seal against sewer gas
B. permit cleaning out the drain line
C. permit the recovery of valuables accidentally dropped into the fixture
D. permit the making of tests on the drain line

26. The one of the following which contains exactly 10 board feet is a board 10 feet long, inches wide, _______ inch(es) thick. 26.____

A. 24; 1
B. 12; 2
C. 12; 1
D. 10; 1

27. Short pieces of pipe threaded on both ends are called 27.____

A. nipples
B. couplings
C. bushing
D. sleeves

28. The unit of electrical capacitance is the 28.____

A. ampere
B. farad
C. henry
D. cycle

29. As used in the electrical industry, BX means 29.____

A. best grade of electrical wire
B. type B extension wire
C. metal greenfield
D. insulated wires in flexible metal tubing

Questions 30-33.

DIRECTIONS: Questions 30 to 33, inclusive, are to be answered in accordance with the following paragraph.

One of the categories of nuisance is a chemical one and relates to the dissolved oxygen of the watercourse. The presence in sewage and industrial wastes of materials capable of undergoing biochemical oxidation and resulting in reduction of oxygen in the watercourse leads to a partial or complete depletion of this oxygen. This, in turn, leads to the subsequent production of malodorous products of decomposition, to the destruction of aquatic plant life and major fish life and to conditions offensive to sight and smell.

30 The word *malodorous* as used in the above paragraph means MOST NEARLY 30.____

A. fragrant B. fetid C. wholesome D. redolent

31 From the above paragraph, because of pollution, the amount of dissolved oxygen in the waterways is 31.____

A. released B. multiplied C. lessened D. saturated

32. The word *categories* as used in the above paragraph means MOST NEARLY 32.____

A. divisions B. clubs C. symbols D. products

33. The word *offensive* as used in the above paragraph means MOST NEARLY 33.____

A. pliable B. complaint
C. deferential D. disagreeable

34. The terminal voltage of 5 dry cells connected in a series is _______ of one(each) cell. 34.____

A. 1/5 the voltage B. the same as the voltage
C. 5 times the voltage D. determined by the current

35. If a 15 ampere fuse blows out and blows out again after inserting a new fuse, it is BEST to 35.____

A. replace it with a 10 ampere fustat
B. replace it with two 10 ampere fuses connected in series
C. replace it with a 20 ampere fuse
D. have the circuit checked to find the trouble

36. Ordinary soft solder is a mixture of lead and 36.____

A. sulphur B. brass C. zinc D. tin

37. The electrolyte used in the ordinary flashlight-type dry cell is 37.____

A. calcium chloride B. ammonium chloride
C. manganese dioxide D. sulfuric acid

38. An electrical transformer is an electrical device used primarily to 38.____

A. raise or lower A.C. voltages
B. change the frequency of alternating current
C. rectify currents from A.C. to D.C.
D. change currents from D.C. to A.C.

39. Of the following, the MAIN reason for the grounding of electrical equipment and circuits is to 39.____

A. save power
B. increase the voltage
C. protect personnel from electric shock
D. prevent serious short circuits

40. In order to properly ground portable electric hand tools, it is USUALLY necessary to use a 40.____

A. solenoid
B. circuit breaker
C. fuse
D. three prong plug

41. The current in a simple electrical circuit can be calculated by dividing the voltage by the resistance in ohms. Assume that the resistance of a certain circuit is 60 ohms and its voltage is 120 volts, 60 cycle A.C. The current in this circuit will be MOST NEARLY _______ ampere(s). 41.____

A. 1/2
B. 2
C. 1
D. 30

42. The one of the following which is the MOST common type of motor that may be used with an A.C. or D.C. source of supply is the _______ motor. 42.____

A. shunt
B. squirrel cage
C. compound
D. series

43. The electrolyte in the ordinary storage battery is 43.____

A. nitric acid
B. sulphuric acid
C. manganese dioxide
D. ammonium chloride

44. The one of the following terms which is used in expressing the rating of a storage battery is 44.____

A. ampere-hours
B. amperes
C. volt-ampere
D. watt-hours

45. The size of the SMALLEST graduation on the ordinary 6-foot folding rule is *usually* 45.____

A. 1/8"
B. 1/16"
C. 1/32"
D. 1/64"

46. A given saw has 8 points per inch. This saw is PROBABLY a _______ saw. 46.____

A. cross-cut
B. hack
C. veneer
D. back

47. Assume that it takes 6 men 8 days to do a certain job. Working at the same same speed, the number of days that it will take 4 men to do this job is 47.____

A. 9 B. 10 C. 12 D. 14

48. The sum of 3 5/8 + 4 1/4 + 6 1/2 + 7 1/8 is 48.____

A. 20 7/8 B. 21 1/4 C. 21 1/2 D. 22 1/8

49. The fraction which is equal to .0625 is 49.____

A. 1/64 B. 3/64 C. 1/16 D. 5/8

50. The volume, in cubic feet, of a rectangular coal bin 8 ft. long by 5 ft. wide by 7 ft. high is MOST NEARLY 50.____

A. 40 B. 56 C. 186 D. 280

KEY (CORRECT ANSWERS)

1. A	11. C	21. A	31. C	41. B
2. D	12. C	22. C	32. A	42. D
3. C	13. A	23. B	33. D	43. B
4. B	14. B	24. D	34. C	44. A
5. C	15. D	25. A	35. D	45. B
6. B	16. A	26. C	36. D	46. A
7. A	17. B	27. A	37. B	47. C
8. B	18. A	28. B	38. A	48. C
9. B	19. B	29. D	39. C	49. C
10. C	20. D	30. B	40. D	50. D

TEST 2

DIRECTIONS: Each question or incomplete statement is followed by several suggested answers or completions. Select the one that BEST answers the question or completes the statement. *PRINT THE LETTER OF THE CORRECT ANSWER IN THE SPACE AT THE RIGHT.*

Questions 1-7

DIRECTIONS: Questions 1 to 7, inclusive, are to be answered in accordance with the following information.

At sea level, the atmosphere can exert a pressure of 14.7 pounds per square inch. This pressure is capable of sustaining a column of water having a height equal to 14.7 pounds, multiplied by 2.304 (the height of water in feet which will exert one pound per square inch pressure). No pump built can produce a perfect vacuum. The atmospheric pressure exerting its force on the surface of the water from which suction is being taken forces the water up through the suction to the pump. From this, it is evident that the maximum height which a water pump of this type can lift water is determined ultimately by the atmospheric pressure. The tightness of the pump and its ability to create a vacuum also have a bearing.

1. The meaning of the word *vacuum* as used in the above article is a 1.____

   A. space entirely devoid of matter
   B. sealed tube filled with gas
   C. bottle-shaped vessel with a double wall
   D. cleaning device

2. With reference to the above article, if a pump could produce a perfect vacuum, the MAXIMUM height, in feet, that it could lift water at sea level is MOST NEARLY 2.____

   A. 33.9 B. 29.4 C. 23.3 D. 14.7

3. With reference to the above article, a column of water having a height of 4.6 feet at sea level will exert a pressure of MOST NEARLY _______ pound(s) per square inch. 3.____

   A. 3 B. 2 C. 1 D. 1/2

4. The word *atmosphere* as used in the above article means 4.____

   A. the pull of gravity
   B. perfect vacuum
   C. the whole mass of air surrounding the earth
   D. the weight of water at sea level

5. The word *bearing* as used in the above article means MOST NEARLY 5.____

   A. direction
   B. connection
   C. divergence
   D. convergence

6. The word *evident* as used in the above article means MOST NEARLY 6.____

A. disconcerting B. obscure C. equivocal D. manifest

7. The word *maximum* as used in the above article means MOST NEARLY 7.____

A. best B. median C. adjacent D. greatest

8. Assume that a car travels at a constant speed of 36 miles per hour. The speed of this car, in feet per second, is MOST NEARLY (one mile equals 5,280 ft.) 8.____

A. 3 B. 24.6 C. 52.8 D. 879.8

9. If one-third of a 19-foot length of lumber is cut off, the length of the remaining piece will measure APPROXIMATELY 9.____

A. 8'8" B. 9'8" C. 12'8" D. 13'8"

10. The circumference of a circle having a diameter of 10" is MOST NEARLY _____ inches. 10.____

A. 3.14 B. 18.72 C. 24.96 D. 31.4

11. Assume that in the purchase of paint, the seller quotes a discount of 10%. If the price per gallon is $6.35, the actual payment in dollars per gallon is MOST NEARLY 11.____

A. $5.72 B. $5.95 C. $6.25 D. $6.50

12. On a 1" bolt that has 10 threads per inch, if the nut is turned 6 complete turns, the distance, in inches, that the nut will move along the bolt is MOST NEARLY 12.____

A. .3 B. .6 C. .9 D. 1

13. Assume that at one end of a 6-inch horizontal line, an 8-inch vertical line is drawn at right angles to the horizontal line. The length, in inches, between the ends of the two lines is MOST NEARLY 13.____

A. 6 B. 8 C. 10 D. 12

Questions 14-21.

DIRECTIONS: Questions 14 to 21, inclusive, are to be answered in accordance with the following information.

In his 2012 annual report to the Mayor, the Public Works Commissioner stated that the city's basic water pollution control program begun in 1996 and costing $425 million so far would be completed in five or six years at a cost of $275 million more. However, he said, the city must spend an additional $175 million more on its marginal pollution control program to protect present and proposed beaches. Under the basic program, the city will have eliminated the last major discharges of raw sewage into the harbor. Over 800 million gallons, two thirds of the city's spent water each day, is now treated at 12 plants, to which six new plants will be added, enabling the city to treat the estimated 1.8 billion gallons that will be discharged daily in 2030. The department had about $200 million worth of municipal construction under way in 2012, and completed $85.5 millions' worth.

14. According to the above, the city will add _______ new plants. 14._____

A. 18 B. 12 C. 6 D. 4

15. The amount of municipal construction under way in 2012 was _______ million. 15._____

A. $85.5 B. $175 C. $200 D. $425

16. It is estimated that in 2030, the city will treat daily _______ gallons. 16._____

A. 700 million
B. 800 million
C. 900 million
D. 1.8 billion

17. According to the above article, the total cost of the water pollution program begun in 1996 will be _______ million. 17._____

A. $275 B. $425 C. $700 D. $815

18. According to the above article, to protect present and proposed beaches, the city must spend an additional _______ million. 18._____

A. $175 B. $275 C. $425 D. $450

19. The above article concerns the statements of the Commissioner of Public Works in his _______ annual report to the Mayor. 19._____

A. 1996 B. 2002 C. 2012 D. 2013

20. The word *discharged* as used in the above article means MOST NEARLY 20._____

A. emitted B. erased C. refuted D. repelled

21. The word *pollution* as used in the above article means MOST NEARLY 21._____

A. condensation
B. purification
C. contamination
D. distillation

22. A tool commonly used to cut off the head of a rivet is a 22._____

A. cold chisel B. cape chisel C. band saw D. file

23. A metal washer is MOST often used with a _______ screw. 23._____

A. wood B. lag C. hand D. machine

24. A good safety rule to follow is that water should NOT be used to extinguish fires in or around electrical apparatus. Of the following, the PRIMARY reason for this is that water 24._____

A. will damage the insulation
B. will corrode the electrical conductors
C. may cause the circuit fuse to blow
D. may conduct electric current and cause a shock hazard

25. One should be extremely careful to keep open flames and sparks away from storage batteries when they are being charged because the 25.____

A. sulphate given off during this operation is highly flammable
B. hydrogen given off during this operation is highly flammable
C. oxygen given off during this operation is extremely flammable
D. static electricity of the battery may cause combustion

26. A good safety rule to follow is that an electric hand tool, such as a portable electric drill, should never be lifted or carried by its service cord. 26.____
Of the following, the PRIMARY reason for this rule is that the

A. tool might swing and be damaged by striking some hard object
B. cord might be pulled off its terminals and become short circuited
C. tool may slip out of the hand as it is hard to get a good grip on a slick rubber cord
D. rubber covering of the cord might overstretch

27. When a man is working on a 15-foot ladder with its top placed against a wall, the MAXIMUM safe distance that he may reach out to one side of the ladder is 27.____

A. as far out as he can reach lifting one foot off the rung for balance
B. as far out as he can reach without bending his body more than 45 from the vertical
C. one-third the length of the ladder
D. as far out as his arm's length

28. When NOT in use, oily waste rags should be stored in 28.____

A. water-tight oak barrels
B. open metal containers
C. sealed cardboard boxes
D. self-closing, metal containers

29. Assume that one of your co-workers has suffered an electric shock. 29.____
Artificial respiration should be started on him immediately if he is

A. unconscious and breathing
B. conscious and in a daze
C. unconscious and not breathing
D. conscious and badly burned

30. Assume that the top of a 12-foot portable straight ladder is placed against a wall but is not held by a man or fastened in any way. In order to be safe, the ladder should be placed so that the distance from the wall to the foot of the ladder is 30.____

A. not over 3 feet
B. not over 4 feet
C. at least 4 feet
D. at least 5 feet

31. Of the following, the one which is an acceptable method of caring for wooden ladders is to 31.____

A. coat the ladder with clear shellac
B. paint the ladder with red lead followed by a second coat of the desired color
C. paint the ladder with a coat of paint of the desired color
D. apply a sealer coat before painting with a second coat of the desired color

32. The MOST important safety precaution to follow when using an electric drill press is to 32.____

A. wear safety shoes
B. drill at a slow speed
C. use plenty of cutting oil
D. clamp the work firmly

33. The proper method of lifting heavy objects is to stand 33.____

A. far enough away from the load so that, with knees bent, the back is at an angle of 45, then lift by straightening the back
B. close to the load, with feet solidly placed and slightly apart and knees bent; then lift by straightening the legs, keeping the back as nearly vertical as possible
C. close to the load, with feet solidly placed and far apart, knees bent; then lift by straightening the legs, keeping the body at an angle of 30°
D. far away from the load, with knees bent and the back at an angle of 45°, then lift by straightening the knees and slowly straightening the back

34. An oilstone is often made of 34.____

A. silicon
B. carborundum
C. tungsten
D. emery

35. To draw a circle, you should use a(n) 35.____

A. compass
B. caliper
C. awl
D. gage

36. A *mushroomed* head is a common defect of a 36.____

A. rivet
B. hammer
C. chisel
D. screwdriver

37. The tool USUALLY used to drive a lag screw is a(n) 37.____

A. open end wrench
B. Stillson wrench
C. screwdriver
D. Allen wrench

38. It is BEST to lubricate machinery 38.____

A. whenever you feel the oil is running low
B. only if the machinery needs it
C. when the machine begins to vibrate
D. on a regular schedule

39. When repairing machinery that is to be reassembled, punch marks are often placed on parts that are next to each other. 39.____
The reason for this is to

A. make sure you assemble the pieces in proper order
B. make it easier to line up the parts in proper position
C. keep count of the number of pieces that belong to this machine
D. provide a stop so that parts cannot be assembled too tightly

Questions 40-46.

DIRECTIONS: Questions 40 to 46, inclusive, are to be answered in accordance with the following paragraph.

At 2:30 P.M. on Monday, October 25, Mr. Paul Jones, a newly appointed Sewage Treatment Worker, started on a routine inspectional tour of the settling tanks and other sewage treatment works installations of the plant to which he was assigned. At 2:33 P.M., Mr. Jones discovered a co-worker, Mr. James P. Brown, lying unconscious on the ground. Mr. Jones quickly reported the facts to his immediate superior, Mr. Jack Rota, who immediately telephoned for an ambulance. Mr. Rota then rushed to the site and placed a heavy woolen blanket over the victim. Mr. Brown was taken to the Ave. H hospital by an ambulance driven by Mr. Dave Smith, which arrived at the sewage disposal plant at 3:02 P.M. Patrolman Robert Daly, badge number 12520, had arrived before the ambulance and recorded all the details of the incident, including the statements of Mr. Jones, Mr. Rota, and Mr. Nick Nespola, a Stationary Engineer (Electric), who stated that he saw the victim when he fell to the ground.

40. The time which elapsed between the start of the sewage treatment worker's routine inspection and the arrival of the ambulance was MOST NEARLY _______ minutes. 40.____

A. 3 B. 28 C. 29 D. 32

41. The name of the sewage treatment worker's immediate superior was 41.____

A. James P. Brown
B. Jack Rota
C. Paul Jones
D. Robert Daly

42. The name of the patrolman was 42.____

A. James P. Brown
B. Jack Rota
C. Paul Jones
D. Robert Daly

43. Referring to the above, the incident occurred on 43.____

A. Monday, Oct. 25
B. Monday, October 26
C. Tuesday, Oct. 25
D. Tuesday, October 26

44. The victim was found at exactly 44.____

A. 2:30 A.M. B. 2:33 P.M. C. 2:33 A.M. D. 2:30 P.M.

45. The sewage treatment worker's name was 45.____

A. James P. Brown
B. Jack Rota
C. Paul Jones
D. Dave Smith

46. The man named Nick Nespola was the 46.____

A. Stationary Engineer (Electric)
B. patrolman
C. victim
D. ambulance driver

47. When sharpening a tool on a grindstone, the tool is often dipped in water. The MAIN reason for this is to 47.____

A. prevent overheating of the tool
B. lubricate the grindstone
C. produce a sharper edge on the tool
D. anneal the tool

48. It is BEST to use a screwdriver having a square shank 48.____

A. when clearance is limited
B. on sheet metal screws
C. on small screws
D. where a wrench is to be used to help turn the screwdriver

49. Brass liners are often placed over the jaws of a bench vise to 49.____

A. grip the work better
B. prevent damage to the work
C. protect the vise
D. make it easier to adjust the work

50. Other than the bulb, the part of a fluorescent light that must be changed MOST often as it wears is the 50.____

A. switch
B. ballast
C. control
D. starter

KEY (CORRECT ANSWERS)

1.	A	11.	A	21.	C	31.	A	41.	B
2.	A	12.	B	22.	A	32.	D	42.	D
3.	B	13.	C	23.	D	33.	B	43.	A
4.	C	14.	C	24.	D	34.	B	44.	B
5.	B	15.	C	25.	B	35.	A	45.	C
6.	D	16.	D	26.	B	36.	C	46.	A
7.	D	17.	C	27.	D	37.	A	47.	A
8.	C	18.	A	28.	D	38.	D	48.	D
9.	C	19.	C	29.	C	39.	B	49.	B
10.	D	20.	A	30.	A	40.	D	50.	D

EXAMINATION SECTION
TEST 1

DIRECTIONS: Each question or incomplete statement is followed by several suggested answers or completions. Select the one that BEST answers the question or completes the statement. *PRINT THE LETTER OF THE CORRECT ANSWER IN THE SPACE AT THE RIGHT.*

1. The flow of sewage into the treatment plant is USUALLY controlled by a 1.____

   A. gate valve
   B. sluice gate
   C. tainter gate
   D. parshall gate

2. Regulator gates USUALLY close when the sewage in the interceptor sewer reaches a predetermined 2.____

   A. velocity
   B. pressure
   C. temperature
   D. elevation

3. A bar screen serves the same purpose as a 3.____

   A. filter
   B. grit collector
   C. trash rack
   D. sluice gate

4. Material that is removed from the sewage by the fine screen is MOST frequently 4.____

   A. blown by compressed air to the grit storage tank
   B. ground up and returned to the sewage
   C. burnt as fuel for the plant
   D. dried in the sludge drying beds and then disposed at sea

5. The one of the following pieces of equipment that is operated in conjunction with air pressure is a(n) 5.____

   A. centrifugal pump
   B. venturi
   C. ejector
   D. sump pump

6. One of the methods used to prime a centrifugal pump is to 6.____

   A. raise the air pressure in the pump
   B. bleed the suction line
   C. apply a vacuum to the pump
   D. open the suction valve

7. The one of the following types of pumps MOST frequently used to pump thickened sludge is the _______ type. 7.____

   A. ejector
   B. centrifugal
   C. gear
   D. piston

8. A plant is called an *Activated Sludge Plant* when the 8.____

   A. thickened sludge can be used as fertilizer
   B. gases from the sludge digestion tanks are burnt as fuel
   C. sludge must be dried before being disposed
   D. partly treated sewage is mixed with sludge

9. Where a digester tank has either a floating or a rising cover, it is made airtight by means of a(n) 9.____

A. water seal
B. sliding rubber gasket
C. leather *bellows*
D. oiled steel ring

10. Grease and fats are USUALLY removed from the sewage by 10.____

A. skimming the liquid in the sedimentation tanks
B. pumping the liquid from the sump in the grit chamber
C. decanting the liquid in the digestion tank
D. backwashing the fine screens

11. The type of plant in which *flocculation* MOST frequently occurs is the _______ plant. 11.____

A. aerated sludge
B. chemical precipitation
C. plain screening
D. filtration

12. Chlorine leaks are BEST detected by use of 12.____

A. orthotoludin
B. litmus paper
C. ammonia
D. copperas

13. Settling tanks operate effectively by _______ the sewage. 13.____

A. slowing the speed of
B. increasing the speed of
C. changing the direction of flow of
D. adding air to

14. A *venturi* is used in a sewage treatment plant in order to 14.____

A. clean the diffusers
B. control the amount of sewage in the wet well
C. measure the flow of sewage
D. reduce the pressure of the gases used as fuel

15. Sludge tanks are USUALLY heated by means of 15.____

A. forced warm air
B. hot water
C. radiant heat
D. electric coils

16. Chlorine is USUALLY added to sewage by 16.____

A. adding the gas directly to the sewage
B. mixing a small quantity of the sewage with the chlorine and then adding the mixture to the main body of sewage
C. mixing the chlorine with water, and then adding the mixture to the sewage
D. combining the gas with the air used in the aeration tank

17. The MAIN reason for defective operation of an aeration tank is that 17.____

A. sewage flow is too slow
B. of clogged diffuser plates
C. tank temperature is too low
D. too much air is supplied

18. The type of pump that seldom requires a relief valve is MOST likely a ______ pump. 18.____

A. reciprocating
B. gear
C. piston
D. centrifugal

19. The MAIN purpose of a foot valve in a centrifugal pump is to 19.____

A. prevent the liquid from flowing back down the suction line
B. equalize the pressure on both sides of the pump
C. make it easier to prime the pump
D. block passage of material that is too large to pump

20. The MAIN reason for lubricating machinery is to 20.____

A. lower operating temperature
B. keep down noise and vibration
C. reduce friction
D. lower cost of operation

21. The one of the following items that has the LOWEST viscosity is 21.____

A. cup grease
B. kerosene
C. #10 oil
D. #40 oil

22. The one of the following statements that is MOST NEARLY correct is: 22.____

A. High speed machinery is most frequently lubricated by grease
B. For most applications, either grease or oil can be used
C. When in doubt, it is best to use the heavier of two grades of oil available
D. Oil becomes *thinner* as the operating temperature increases

23. The function of a circuit breaker is MOST similar to that of a 23.____

A. switch B. fuse C. rheostat D. transformer

24. Noisy operation of a motor is MOST frequently caused by 24.____

A. a shorted armature
B. over-voltage
C. worn bearings
D. a grounded casing

25. Consumption of electrical energy is registered on a(n) 25.____

A. volt-ohm meter
B. ammeter
C. watt-hour meter
D. ohm meter

26. A dirty commutator is BEST cleaned by using 26.____

A. sandpaper
B. soap and water
C. emery cloth
D. kerosene

27. The one of the following items that is MOST frequently used to prevent an electric motor from being overloaded is a 27.____

A. warning signal
B. governor
C. thermal cut-out
D. rheostat

28. The one of the following metals that is MOST commonly used for outboard bearings is 28.____

A. zinc B. brass C. magnesium D. babbit

29. The use of pipe joint compound when making up a screwed joint results in a watertight joint and also 29.___

A. cleans the threads
B. makes the joint hard
C. lubricates the threads
D. prevents cross threading

30. A pipe is generally threaded by using a 30.___

A. die
B. tap
C. yoke
D. swedge

31. The type of motor that MOST frequently does NOT use brushes is the 31.___

A. universal type
B. series wound D.C. motor
C. synchronous motor
D. induction motor

32. Compressed air can be used to clean generator and motor windings provided the air is 32.___

A. heated
B. blown at a high velocity
C. used at a pressure of at least 90 lbs./sq.in.
D. dry

33. The use of a cold chisel with a *mushroomed* head is 33.___

A. *good,* because the mushrooming cushions the blow
B. *bad,* because the head cannot be hit squarely
C. *good,* because there is more area on the head to hit
D. *bad,* because chips might fly from the head

34. After brass or black iron pipe has been cut, it should be 34.___

A. counterbored
B. reamed
C. countersunk
D. squared

35. The one of the following that is used to change the speed of certain types of electric motors is a (n) 35.___

A. commutator
B. brush
C. rheostat
D. armature

36. The type of pipe that is MOST frequently made with bell and spigot ends is 36.___

A. brass
B. steel
C. cast iron
D. transite

37. The one of the following that is used to connect two pipes together in a straight line is a 37.___

A. manifold
B. divider
C. band
D. union

38. The difference between a stud and a bolt is that the stud has 38.___

A. a finer thread
B. no head
C. a round head
D. a coarser thread

39. A set screw is often used to 39.____

A. bolt two pieces of flanged pipe together
B. screw together matching parts in a motor casing
C. clamp a piece to a work table
D. secure a pulley to a shaft

40. Soft jaw inserts sometimes used to protect the surface of a piece of metal that is held in a vise are MOST frequently made of 40.____

A. zinc B. tin C. brass D. pewter

41. The BEST type of wrench to use on a large square nut is a ______ wrench. 41.____

A. monkey B. spanner C. stillson D. spintite

42. The BEST method of cleaning files is to use a 42.____

A. file card B. knife
C. scriber D. fibre brush

43. The BEST lubricant to use when cutting threads on steel pipe is ______ oil. 43.____

A. pike B. penetrating
C. lard D. coal

44. The BEST type of valve to use to control the flow of liquid to a delicate gauge is a ______ valve. 44.____

A. gate B. needle C. globe D. check

45. Water hammer is caused MAINLY by 45.____

A. pumping sewage to too high a head
B. interrupting the flow of sewage too rapidly
C. debris floating in the sewage
D. excessive corrosion in the pipes

46. Suppose a centrifugal pump is pumping less sewage than it is capable of handling. Of the following, the one that is NOT a possible reason for this is that the 46.____

A. speed of pump is too slow
B. pump is not properly primed
C. stuffing box packing is defective
D. suction line is partly clogged

47. The one of the following types of pumps that will give a smooth continuous flow of liquid rather than a pulsating flow is the ______ type. 47.____

A. reciprocating B. rotary
C. gear D. centrifugal

48. For pumping against a very high head, the BEST type of pump to use is a ______ type. 48.____

A. reciprocating B. propeller
C. mixed flow D. centrifugal

49. To increase the volume of delivery of a reciprocating pump, USUALLY the 49.____

A. angle of the impeller is increased
B. inlet valve is opened wider
C. piston stroke is lengthened
D. discharge valve is opened wider

50. The capacity of a pump is MOST frequently expressed in 50.____

A. cubic feet per day
B. gallons per day
C. cubic feet per minute
D. gallons per minute

KEY (CORRECT ANSWERS)

1. B	11. B	21. B	31. D	41. A
2. D	12. C	22. D	32. D	42. A
3. C	13. A	23. B	33. D	43. C
4. B	14. C	24. C	34. B	44. B
5. C	15. B	25. C	35. C	45. B
6. C	16. C	26. A	36. C	46. B
7. D	17. B	27. C	37. D	47. D
8. D	18. D	28. D	38. B	48. A
9. A	19. A	29. C	39. D	49. C
10. A	20. C	30. A	40. C	50. D

TEST 2

DIRECTIONS: Each question or incomplete statement is followed by several suggested answers or completions. Select the one that BEST answers the question or completes the statement. *PRINT THE LETTER OF THE CORRECT ANSWER IN THE SPACE AT THE RIGHT.*

1. The sum of the following dimensions: 1 5/8, 2 1/4, 4 1/16, 3 3/16, is 1.____

   A. 10 15/16 B. 11 C. 11 1/8 D. 11 1/4

2. Assume that six men, working together at the same rate of speed, can complete a certain job in 3 hours. 2.____
   If, however, there were only four men available to do this job, and they all worked at the same rate of speed, to complete this job would take MOST NEARLY ______ hours.

   A. 4 1/4 B. 4 1/2 C. 4 3/4 D. 5

3. Due to unforeseen difficulties, a job which would normally take 17 hours to complete was actually completed in 21 hours. 3.____
   This represents a percent increase over the normal time of MOST NEARLY

   A. 19% B. 2.4% C. 24% D. 124%

4. The veteran should approach the problem of safety with the idea that 4.____

   A. there will always be accidents
   B. most accidents can be prevented
   C. the best method of preventing accidents is to post safety rules for the men to follow
   D. punishing the man with the worst accident record will reduce the number of accidents occurring

5. The one of the following that is NOT a common cause of accidents occurring when working around machinery is 5.____

   A. wearing loose clothing
   B. wearing gloves
   C. having insufficient illumination
   D. having slippery floors

Questions 6-8.

DIRECTIONS: Questions 6 through 8, inclusive, are to be answered in accordance with the following information.

A certain job requires 4 men working the number of hours and at the salary rate indicated in the accompanying table.

Name	No. of Hours	Salary/Hr.
Brown	30	$15.00
Jones	22	$19.50
Walter	40	$10.50
Thomas	25	$17.22

6. According to the above table, the salary received by Thomas on this job is MOST NEARLY 6.___

A. $426.00 B. $427.50 C. $429.00 D. $430.50

7. According to the above table, the man who received the MOST wages chargeable to this job is 7.___

A. Brown B. Jones C. Walter D. Thomas

8. According to the above table, the total amount of wages chargeable to this job is MOST NEARLY 8.___

A. $1,726.50 B. $1,717.50 C. $1,729.50 D. $1,737.50

9. Of the following statements, the one that represents the SAFEST practice in a shop is: Adjustments should be made on. 9.___

A. running machinery only if another man can be assigned to guard the man making the actual adjustment
B. running machinery only if proper protective equipment is worn
C. running machinery only when the machine is grounded
D. machinery only after the machine has been stopped

10. Regarding work performed on electrical circuits, the one of the following that is unsafe is to 10.___

A. use #10 wire instead of #12
B. ground the junction boxes
C. replace a 15 ampere circuit breaker with a 20 ampere one
D. open the main switch before working on the wiring

11. Of the following, the MOST important reason for making detailed reports of all accidents is to 11.___

A. have a record of who to blame for the accident
B. be able to properly assess the cost of the accident
C. reduce the number of *compensation* claims
D. determine the causes of accidents and eliminate future accidents

12. As a veteran sewage treatment worker, you can BEST promote safety in your operations by 12.___

A. carefully investigating and reporting the circumstances of any accident
B. suggesting safer methods of operation
C. training subordinates in proper safety
D. disciplining subordinates who engage in unsafe acts

13. Oil-soaked rags are BEST stored in a 13.___

A. neat pile in a readily accessible corner
B. metal container with a tight cover
C. metal box that has holes for adequate ventilation
D. closet on a shelf above the ground

14. The one of the following actions that is NOT the cause of injury when working with hand tools is 14.____

A. working with defective tools
B. using the wrong tool for the job
C. working too carefully
D. using a tool improperly

15. Artificial respiration is the FIRST action you should take when a man becomes unconscious either as a result of drowning or as a result of 15.____

A. chlorine poisoning
B. electric shock
C. falling
D. clothing catching fire

Questions 16-17.

DIRECTIONS: Questions 16 and 17 should be answered in accordance with the following paragraph.

Sewage treatment plants are designed so that the sewage flow reaches the plant by gravity. In some instances, a small percentage of the sewerage system may be below the planned level of the plant. Economy in construction and other factors may indicate that the raising of that lower portion of the flow by means of pumps, to the desired plant elevation, is more desirable than lowering the plant structure. Some plants operate with this feature.

16. According to the above paragraph, 16.____

A. a small percentage of the sewage reaches the plant by means of gravity
B. all sewage reaches the plant by means of gravity
C. where sewage cannot reach the plant by gravity it is pumped
D. pumping is used so that all sewage can reach the plant

17. According to the above paragraph, the reason that some plants are built above the level of the sewerage system is that 17.____

A. these plants operate more efficiently this way
B. gravity will naturally bring the sewage in at a lower level
C. pumping of the sewage is more expensive
D. these plants are cheaper to build this way

Questions 18-20.

DIRECTIONS: Questions 18 through 20, inclusive, should be answered in accordance with the following paragraph.

Accident proneness is a subject which deserves much move objective and competent study than it has received to date. In discussing accident proneness, it is important to differentiate between the employee who is a "repeater" and one who is truly accident prone. It is obvious that any person put on work of which he knows little without thorough training in safe practice for the work in question will be liable to injury until he does learn the "how" of it. Few workmen left to their own devices will develop adequate safe practices. Therefore, they must be trained. Only those who fail to respond to proper training should be regarded as accident prone. The repeater whose accident record can be explained by a correctible physical defect, by correctible plant or machine hazards, by assignment to work for which he is not suited because of physical deficiencies or special abilities, cannot be fairly called "accident prone."

18. According to the above paragraph, a person is considered accident prone if 18.____

    A. he has accidents regardless of the fact that he has been properly trained
    B. he has many accidents
    C. it is possible for him to have accidents
    D. he works at a job where accidents are possible

19. According to the above paragraph, 19.____

    A. workers will learn the safe way of doing things if left to their own intelligence
    B. most workers must be trained to be safe
    C. a worker who has had more than one accident has not been properly trained
    D. intelligent workers are always safe

20. According to the above paragraph, a person would not be called accident prone if the cause of his accidents was 20.____

    A. a lack of interest in the job
    B. recklessness
    C. a low level of intelligence
    D. eyeglasses that don't fit properly

Questions 21-23.

DIRECTIONS: Questions 21 through 23, inclusive, should be answered in accordance with the following paragraph.

Sharpening a twist drill by hand is a skill that is mastered only after much practice and careful attention to the details. Therefore, whenever possible, use a tool grinder in which the drills can be properly positioned, clamped in place, and set with precision for the various angles. This machine grinding will enable you to sharpen the drills accurately. As a result, they will last longer and will produce more accurate holes.

21. According to the above paragraph, one reason for sharpening drills accurately is that the drills 21.____

    A. can then be used in a hand drill as well as a drill press
    B. will last longer
    C. can then be used by unskilled persons
    D. cost less

22. According to the above paragraph, 22.____

    A. it is easier to sharpen a drill by machine than by hand
    B. drills cannot be sharpened by hand
    C. only a skilled mechanic can learn to sharpen a drill by hand
    D. a good mechanic will learn to sharpen drills by hand

23. As used in the above paragraph, the word *precision* means MOST NEARLY 23.____

    A. accuracy B. ease C. rigidity D. speed

Questions 24-27.

DIRECTIONS: Questions 24 through 27, inclusive, should be answered in accordance with the following paragraph.

Centrifugal pumps have relatively fewer moving parts than reciprocating pumps, and no valves. While reciprocating pumps when new are usually more efficient than centrifugal pumps, the latter retain their efficiency longer. Most rotary pumps are also without valves, but they have closely meshed parts between which high pressures may be set up after they begin to wear. In general, centrifugal pumps can be made much smaller than reciprocating pumps giving the same result. There is an exception, in that positive displacement pumps delivering small volumes at high heads are smaller than equivalent centrifugal pumps. Centrifugal pumps cost less when first purchased than other comparable pumps. The original outlay may be as little as one-third or one-half that of a reciprocating pump suitable for the same purpose.

24. The type of pump NOT mentioned in the above paragraph is the _______ type. 24.____

A. rotary
B. propeller
C. reciprocating
D. centrifugal

25. According to the above paragraph, the type of pump that sometimes has valves and sometimes does NOT is the 25.____

A. rotary
B. propeller
C. reciprocating
D. centrifugal

26. According to the above paragraph, centrifugal pumps are 26.____

A. *always smaller* than reciprocating pumps
B. *smaller* than reciprocating pumps only when designed to deliver small quantities at low pressures
C. *larger* than reciprocating pumps only when designed to deliver small quantities at high pressures
D. *larger* than reciprocating pumps only when designed to deliver large quantities at low pressures

27. The advantage of centrifugal pumps that is NOT mentioned in the above paragraph is that 27.____

A. the centrifugal pump retains its efficiency longer
B. it is impossible to create an excessive pressure when using a centrifugal pump
C. there are fewer parts to wear out in a centifugal pump
D. the centrifugal pump is cheaper

Questions 28-30.

DIRECTIONS: Questions 28 through 30, inclusive, should be answered in accordance with the following paragraph.

Gaskets made of relatively soft materials are placed between the meeting surfaces of hydraulic fittings in order to increase the tightness of the seal. They should be composed of materials that will not be affected by the liquid to be enclosed, nor by the conditions under which the system operates, including maximum pressure and temperature. They should be able to maintain the amount of clearance required between meeting surfaces. One of the materials most widely used in making gaskets is neoprene. Since neoprene is flexible, it is often used in sheet form at points where a gasket must expand enough to allow air to accumulate, as with cover plates on supply tanks. Over a period of time, oil tends to deteriorate the material used in making neoprene gaskets. The condition of the gasket must, therefore, be checked whenever the unit is disassembled. Since neoprene gasket material is soft and flexible, it easily becomes misshapen, scratched or torn. Great care is, therefore, necessary in handling neoprene. Shellac, gasket sealing compounds or "pipe dope" should never be used with sheet neoprene, unless absolutely necessary for satisfactory installation.

28. Of the following, the one that is NOT mentioned in the above paragraph as a requirement for a good gasket material is that the material must be 28.____

    A. cheap
    B. unaffected by heat developed in a system
    C. relatively soft
    D. capable of maintaining required clearances

29. According to the above paragraph, neoprene will be affected by 29.____

    A. air
    B. temperature
    C. pressure
    D. oil

30. According to the above paragraph, care is necessary in handling neoprene because 30.____

    A. its condition must be checked frequently
    B. it tears easily
    C. pipe dope should not be used
    D. it is difficult to use

Questions 31-35.

DIRECTIONS: Questions 31 through 35, inclusive, should be answered in accordance with the following statements and instructions.

Column A below lists defects that often happen to equipment that is used in a sewage disposal plant. Column B shows the equipment that is used in such a plant. In the space at the right next to the number of the defect listed in Column A, select the letter in Column B representing the piece of equipment with which this defect is MOST closely associated.

COLUMN A	COLUMN B	
31. Broken shear pin	A. Centrifugal pump	31.____
32. Worn collector ring	B. Wound rotor motor	32.____
33. Pitted impeller	C. Bar screen	33.____
34. Worn chain	D. Methane-gas engine	34.____
35. Crankpin bearing		35.____

36. It is often said that in selecting a man for a job, dependability is more important than seniority. This is because 36.____

A. it is difficult to judge the amount of work an older man can do
B. an older man will know how to *avoid* work better
C. the dependable man is the man you can count on to do the job as called for
D. the dependable man will require fewer instructions

37. *A man may be conscientious, and yet not be efficient.* This statement MOST likely means that 37.____

A. a man will not be able to do a job properly unless he has special training
B. a man may want to do a job well, but may not know how to go about doing it
C. if a man is efficient, he may not be conscientious
D. the more conscientious a man is the less efficient he will be

38. If you were a senior sewage treatment worker, a good way of building up the morale of men assigned to you would be to 38.____

A. overlook minor infractions of the rules
B. pass the blame for bad assignments to your superiors
C. treat the men fairly
D. cover up for men who have made mistakes in their jobs

39. Threatening your subordinates with penalties for neglect of duty is 39.____

A. *good* practice just to frighten them, even though the penalties will not be inflicted
B. *poor* practice since men should never be threatened
C. *good* practice if the penalty is actually going to be inflicted
D. *poor* practice because men ought to work properly without threats

40. Of the following, the BEST indication that men are dissatisfied with their jobs is that they 40.____

A. offer suggestions on improving operations
B. appoint a grievance committee
C. all join a union
D. frequently leave for other jobs

41. If a senior sewage treatment worker must reprimand one of the men under him, the reprimand should be given 41.____

A. in a loud tone of voice so that the man is properly impressed
B. firmly but quietly
C. the next day when the senior can get the man alone
D. in front of the entire crew so that the rest of the men know what is right

42. If a senior sewage treatment worker is not sure of how a job should be done, he should 42.____

A. make believe he does so that his men will not discover his lack of knowledge
B. get someone else to do the job
C. ask his superior how the job should be done
D. put the job off until he can learn from another crew how it should be done

43. A senior sewage treatment worker makes a mistake, and admits it to his men. This practice is _______, because the men _______. 43._____

A. *good;* will respect him more
B. *poor;* will not trust his judgment anymore
C. *good;* will then learn to check everything he does before wasting time doing jobs improperly
D. *poor;* should not know why a job is being done in the way it is

44. A supervisor can BEST earn the respect of his men by 44._____

A. never criticizing his men
B. taking the blame for all actions of his men
C. defending his men from all criticism, regardless of whether the criticism is deserved or not
D. defending his men from unsupported criticism

45. As a senior sewage treatment worker, you have been ordered by the engineer to do a job in a certain manner which you think is not a good way of doing the job. You should 45._____

A. tell the engineer you should be permitted to do the job in whatever way you feel best
B. avoid doing the job
C. do the job, but tell your men that you are not responsible for the method being used
D. explain your objections to the engineer, but then do the job in whatever manner the engineer decides

46. The MOST important requirement for a good supervisor is to have 46._____

A. physical strength
B. the ability to handle men
C. manual dexterity
D. good appearance

47. A good senior sewage treatment worker should 47._____

A. give all the disagreeable assignments to the laziest worker
B. give all the good assignments to the best worker
C. give disagreeable assignments to those men who have special training for them
D. rotate disagreeable assignments among the men

48. A new sewage treatment worker has been assigned to work under you as a senior. The MOST important information you should get from the new man is 48._____

A. his age
B. the type of work he likes to do
C. his previous experience
D. how well he gets along with other men

49. A member of your crew, who frequently comes to you with unjustified complaints, comes to you again with another complaint. You should 49._____

A. cut the man short and tell him to stop complaining unnecessarily
B. listen to the complaint, but do nothing about it
C. listen to the complaint, and then tell the man the complaint is not justified
D. check the complaint to see if it is justified

50. To insure that the men working under him are doing their work properly, a senior sewage treatment worker should 50.____

A. check their work frequently
B. have the men prepare a written report about the work
C. assign one individual to be responsible for each job
D. keep a record of the supplies they use

KEY (CORRECT ANSWERS)

1. C	11. D	21. B	31. C	41. B
2. B	12. C	22. A	32. B	42. C
3. C	13. B	23. A	33. A	43. A
4. B	14. C	24. B	34. C	44. D
5. B	15. B	25. A	35. D	45. D
6. D	16. B	26. C	36. C	46. B
7. A	17. D	27. B	37. B	47. D
8. C	18. A	28. A	38. C	48. C
9. D	19. B	29. D	39. C	49. D
10. C	20. D	30. B	40. D	50. A

EXAMINATION SECTION
TEST 1

DIRECTIONS: Each question or incomplete statement is followed by several suggested answers or completions. Select the one that *BEST* answers the question or completes the statement. *PRINT THE LETTER OF THE CORRECT ANSWER IN THE SPACE AT THE RIGHT.*

1. In a modification of the conventional activated sludge process known as Modified Aeration, the percentage of returned sludge to the aeration tank is, MOST nearly, 1.____

   A. 10 B. 20 C. 30 D. 40

2. The amount of chlorine, in pounds per million gallons, to produce 0.5 ppm residual in most primary effluents will, *most nearly,* be between 2.____

   A. 10 to 40 B. 50 to 70 C. 100 to 200 D. 300 to 500

3. In a conventional activated sludge treatment plant, air is applied at a rate of, most *nearly,* 3.____

   A. 1 to 1 1/2 cubic feet per gallon of sewage
   B. 3 to 3 1/2 cubic feet per gallon of sewage
   C. 4 to 5 1/2 cubic feet per gallon of sewage
   D. 7 to 7 1/2 cubic feet per gallon of sewage

4. Of the following temperature ranges, the *one* which is the *MOST* efficient for sludge digester operation is 4.____

   A. 45° F and 50° F B. 55° F and 65° F
   C. 70° F and 75° F D. 85° F and 95° F

5. The sewage detention time in an aeration tank using modified aeration is, *most nearly,* 5.____

   A. 2 hours B. 4 hours C. 6 hours D. 8 hours

6. The BTU per cubic foot value of sludge gas from a well established and properly operated digestion tank is, most *nearly,* 6.____

   A. 150 B. 350 C. 450 D. 650

7. BOD is an abbreviation for 7.____

   A. Bacteria Operating Demand
   B. Biosorption Operating Demand
   C. Biochemical Oxygen Demand
   D. Biofilter Oxygen Demand

8. The one of the following that is normally used to control the flow of sewage to the treatment plant from the intercepting sewer is the 8.____

   A. float valve B. sluice gate
   C. gate valve D. regulator gate

9. A sludge gas encountered at sewage treatment plants that is corrosive and damaging to metals is 9.___

A. carbon dioxide
B. ethane
C. nitrogen
D. hydrogen sulphide

10. When sludge is withdrawn from a sludge gas collector tank with a fixed color, a compensating volume of fresh sludge or water or gas must be put into the tank to prevent the development of 10.___

A. leakage
B. positive pressures
C. negative pressures
D. condensation

11. Devices in sewage treatment plants whose function is to break or cut up solids found in sewage are known as 11.___

A. barmimutors
B. diffusers
C. tricklers
D. grinders

12. The sludge treatment process whereby the volume of sludge going to the digester is reduced is known as 12.___

A. thickening
B. elutriation
C. chemical conditioning
D. wet oxidation

13. *Most* of the suspended solids are separated or removed from the sewage by 13.___

A. aeration B. washing C. elutriation D. sedimentation

14. The *one* of the following that is usually operated by compressed air is a 14.___

A. reducer
B. baffle
C. sump pump
D. sewage ejector

15. The *PRIMARY* function of a grit chamber in a sewage treatment plant is to remove 15.___

A. paper B. worms C. gravel D. algae

16. A deep two-storied storage sewage tank with an upper sedimentaton chamber and a lower chamber is known as a _______ tank. 16.___

A. detritus B. imhoff C. septic D. elocculating

17. The *one* of the following which *BEST* characterizes activated sludge is that it is 17.___

A. black in color and has small particles
B. blue in color and has large particles
C. brown in color and has some dissolved oxygen
D. beige in color and has a great amount of dissolved oxygen

18. The *optimum* PH value of the sludge in a digester should be 18.___

A. 10 B. 7 C. 3 D. 2

19. In the Activated Sludge Process, the *one* of the following steps that may be taken to prevent or control sludge bulkings is to 19.___

A. decrease aeration in time and rate
B. chlorinate returned activated sludge

C. increase the solids content carried in aeration tanks;
D. raise the pH value to 7.8

20. In starting a digester unit, the *QUICKEST* results can be obtained by 20.____

A. seeding B. shredding C. dosing D. chlorinating

21. Sludge digestion carried out in the absence of free oxygen is known as 21.____

A. wet oxidation
B. heat drying
C. anaerobic decomposition
D. aerobic decomposition

22. "Frothing" is *MOST* frequently attributable to 22.____

A. short circuiting of aeration tanks
B. septic sewage in primary tank
C. high concentration of fungus
D. detergent compounds in the sewage

23. The process of removing floating grease or scum from the surface of sewage in a tank is called 23.____

A. squeegeeing
B. siphoning
C. skimming
D. sloughing

24. Of the following, the *one* which *BEST* represents a primary treatment device for sewage is the 24.____

A. stabilization pond
B. intermittent sand filter
C. septic tank
D. aeration tank

25. Freshly poured concrete surfaces normally exposed to air should be cured for a minimum period of 25.____

A. 4 days B. 5 days C. 6 days D. 7 days

26. One of your men on the job is injured at a work site and is unconscious. The *BEST* course of action for you to follow is to 26.____

A. give him liquids to drink
B. have him remain in a lying position until medical help arrives
C. immediately move him to the first-aid station
D. attempt to arouse him to consciousness by shaking him

27. The type of portable fire extinguisher that is *MOST* effective in controlling a fire around live electrical equipment is the 27.____

A. foam type
B. soda-acid type
C. carbon-dioxide type
D. water type

28. The hazards of electric shock resulting from operation of a portable electric tool in a damp location can be reduced by 28.____

A. grounding the tool
B. holding the tool with one hand
C. running the tool at low speed
D. using a baffle

29. The *one* of the following that is the *proper* first aid to administer to a conscious person suffering from chlorine inhalation is 29.___

A. an alocholic drink
B. black coffee
C. a pulmotor
D. a cold shower

30. Of the following actions, the *best one* to take *FIRST* after smoke is seen coming from an electric control device is to 30.___

A. shut off the power to it
B. call the main office for advice
C. look for a wiring diagram
D. throw water on it

KEY (CORRECT ANSWERS)

1.	A	16.	B
2.	C	17.	C
3.	A	18.	B
4.	D	19.	B
5.	A	20.	A
6.	D	21.	C
7.	C	22.	D
8.	B	23.	C
9.	D	24.	C
10.	C	25.	D
11.	A	26.	B
12.	A	27.	C
13.	D	28.	A
14.	D	29.	B
15.	C	30.	A

TEST 2

DIRECTIONS: Each question or incomplete statement is followed by several suggested answers or completions. Select the one that *BEST* answers the question or completes the statement. *PRINT THE LETTER OF THE CORRECT ANSWER IN THE SPACE AT THE RIGHT.*

1. Of the following, the *BEST* fastener to use when attaching a pipe support bracket to a concrete wall is a(n) 1._____

   A. toggle bolt B. expansion bolt
   C. carriage bolt D. lag bolt

2. The *MAIN* reason for mixing a "thinner" into paint is to 2._____

   A. *clear up* air bubbles
   B. *stop* the paint from bleeding
   C. *spread* the paint easily
   D. *make* the paint color lighter

3. Schedule 40 pipe is a designation for 3._____

   A. asbestos cement pipe B. steel pipe
   C. transite pipe D. clay pipe

4. The function of a check valve in a pipeline is to 4._____

   A. relieve excessive pressure B. remove air
   C. meter the flow D. prevent reverse flow

5. The device on an electric motor which will prevent overheating is called a 5._____

   A. rheostat B. bus bar
   C. solenoid D. thermal relay

6. The oil recommended for the gear box of a 20-ton sewage plant crane is, *most nearly,* 6._____

   A. SAE 80 B. SAE 120 C. SAE 160 D. SAE 200

7. Where pump ball bearings may be subjected to water washing, the lubricating grease should have a 7._____

   A. white lead base B. red lead base
   C. sodium soap base D. lithum soap base

8. A chlorine leak can normally be detected by 8._____

   A. a lighted candle B. its smell
   C. a dry rag D. an oil-soaked rag

9. The moving wooden planks in a tank used to scrape sludge from the bottom of a tank are known as 9._____

   A. cleats B. flights C. rails D. levers

10. A device with an edge or notch used for measuring liquid flow is known as a 10._____

    A. Parshall Flume B. Plainer Bowlus
    C. Venturi D. Weir

11. The *one* of the following types of pumps that is *WIDELY* used for pumping sewage is 11.___

A. reciprocating B. rotary C. simplex D. centrifugal

12. Prior to starting a newly installed pump, you should 12.___

A. open the motor disconnect switch
B. expose the pump to outside weather conditions
C. turn the shaft by hand to see that it rotates freely
D. disconnect the vent and drain the plugs

13. A maintenance program for a new piece of operating equipment should *BEST* be set up in accordance with the 13.___

A. location of the unit
B. location of personnel
C. manufacturer's recommendations
D. monthly plant capacity

14. The *one* of the following fasteners that has threads at *both* ends is called a 14.___

A. screw B. stud C. blivet D. drift bolt

15. The *one* of the following that is installed between two pipe flanges to seal the connection is called a 15.___

A. sheave B. gasket C. boss D. fillet

16. A wet undigested sludge containing a large amount of grease will *MOST* probably 16.___

A. clog the opening of the filter
B. have no effect on the efficiency of the filters
C. cause rapid deterioration of the filter
D. cause the filter to shrink and snap

17. The floating cover for a sludge gas storage tank is kept under a gauge pressure of, *most nearly,* 17.___

A. 0 to 2 ounces B. 3 to 5 ounces
C. 6 to 9 ounces D. 10 to 12 ounces

18. The tool that is used to remove the burrs from the end of 1/2" diameter steel pipe after cutting it with a pipe cutter is known as a 18.___

A. bit B. reamer C. tap D. caliper

19. Of the following common obstructions found in sewer lines, the *one* that occurs *MOST* frequently is 19.___

A. roots B. debris C. grease D. grit

20. The *one* of the following that is the *MAIN* reason for putting orders in writing is to 20.___

A. protect the person who receives it
B. protect the person who prepared the order
C. make it easier to check mistakes
D. protect the agency should something unforeseen occur

21. For records to provide an essential basis for future changes or expansions of the sewage treatment plant, the records must be 21.____

A. accurate
B. lengthy
C. detailed in ink
D. hand-written in pencil

22. The volume, in cubic feet, of a slab of concrete that is 5'-0" wide, 6'-0" long, and 0'-6" in depth is, *most nearly,* 22.____

A. 15.0 B. 13.5 C. 12.0 D. 10.5

23. The sum of the following pipe lengths, 22 1/8", 7 3/4", 19 7/16", and 4 3 5/8", is: 23.____

A. 91 7/8" B. 92 1/16" C. 92 1/4" D. 92 15/16"

24. The area in square feet of a plant floor that is 42 feet wide and 75 feet long is 24.____

A. 3150 B. 3100 C. 3075 D. 2760

25. Of the following types of gauges, the *one* that indicates pressure above and below atmospheric pressures is known as a 25.____

A. pressure B. vacuum C. Bourdan D. compound

26. A U-tube manometer is used to measure 26.____

A. deflection B. height C. radiation D. pressure

27. If an air-conditioning unit shorted out and caught fire, the *BEST* fire extinguisher to use would be a ______ extinguisher. 27.____

A. water
B. foam
C. carbon dioxide
D. soda acid

28. Of the following, the *best* action to take to help someone whose eyes have been splashed with lye is to *FIRST* 28.____

A. wash out the eyes with clean water
B. wash out the eyes with a salt water solution
C. apply a sterile dressing over the eyes
D. do nothing to the eyes, but telephone for medical help

Questions 29-30.

DIRECTIONS: Questions numbered 29 and 30 are to be answered in accordance with the information given in the following paragraph:

A sludge lagoon is an excavated area in which digested sludge is desired. Lagoon depths vary from six to eight feet. There are no established criteria for the required capacity of a lagoon. The sludge moisture content is reduced by evaporation and drainage. Volume reduction is slow, especially in cold and rainy weather. Weather and soil conditions affect concentration. The drying period ranges from a period of several months to several years. After the sludge drying period has ended, a bulldozer or tractor can be used to remove the sludge. The dried sludge can be used for fill of low ground. A filled dried lagoon can be developed into a lawn. Lagoons can be used for emergency storage when the sludge beds are full. Lagoons are popular because they are inexpensive to build and operate. They have a disadvantage of being unsightly. A hazard to lagoon operation is the possibility of draining partly digested sludge to the lagoon that creates a fly and odor nuisance.

29. In accordance with the given paragraph, sludge lagoons have the *disadvantage* of being 29.____

A. unsightly
B. too deep
C. concentrated
D. wet

30. In accordance with the given paragraph, moisture content is *reduced* by 30.____

A. digestion
B. evaporation
C. oxidation
D. removal

KEY (CORRECT ANSWERS)

1.	B	16.	A
2.	C	17.	B
3.	B	18.	B
4.	D	19.	A
5.	D	20.	B
6.	B	21.	A
7.	D	22.	A
8.	B	23.	D
9.	B	24.	A
10.	D	25.	D
11.	D	26.	D
12.	C	27.	C
13.	C	28.	A
14.	B	29.	A
15.	A	30.	B

EXAMINATION SECTION
TEST 1

DIRECTIONS: Each question or incomplete statement is followed by several suggested answers or completions. Select the one that BEST answers the question or completes the statement. *PRINT THE LETTER OF THE CORRECT ANSWER IN THE SPACE AT THE RIGHT.*

1. To check for the entrance of toxic wastes into a treatment plant, each of the following may be reliably observed as indicators EXCEPT 1.____

   A. changes in color of incoming wastewater
   B. waste recording equipment
   C. odors
   D. bulking of sludge in the clarifier

2. An increase in ______ could cause a demand for more oxygen in an aeration tank. 2.____

   A. inert or inorganic wastes
   B. pH
   C. toxic substances
   D. microorganisms

3. Chlorine may be added for hydrogen sulfide control in the 3.____

   A. collection lines
   B. aeration tank
   C. plant effluent
   D. trickling filter

4. The range of typical carrying capacities, in gallons per minute, of intermediate pumping stations is 4.____

   A. less than 600
   B. 200-700
   C. 100-1,600
   D. 700-10,000

5. A low sulfanator injector vacuum reading could be caused by 5.____

   A. missing gasket
   B. high back pressure
   C. high-volume injector flow
   D. wrong orifice

6. Before starting a rotating biological contactor process, each of the following should be checked EXCEPT 6.____

   A. lubrication
   B. biomass
   C. clearance
   D. tightness

7. The capacity for water or wastewater to neutralize acids is expressed in terms of 7.____

   A. pH
   B. oxygen demand
   C. alkalinity
   D. acidity

8. Which of the following is NOT one of the available methods for determining stormwater flow for the purpose of storm sewer design? 8.____

A. Rainfall and runoff correlation studies
B. Inlet method
C. Hydrograph method
D. Outlet method

9. What is the term for the accumulation of residue that appears on trickling filters and must be removed periodically? 9.____

A. Sludges B. Slurries C. Slugs D. Sloughings

10. A sludge containing a high number of living organisms is referred to as 10.____

A. raw B. activated C. primary D. toxic

11. Which of the following is NOT a plant location where liquid mixing is commonly practiced? 11.____

A. Ponds
B. Hydraulic jumps in open channels
C. Pipelines
D. Venturi flumes

12. Which of the following industries releases primarily inorganic wastes in its effluent? 12.____

A. Paper B. Petroleum
C. Gravel washing D. Dairy

13. Which of the following collection system variables could upset a plant's activated sludge process? 13.____

A. Discharge by industrial cleaning operations
B. Chlorination of return sludge flows
C. Decreases in influent flows
D. Recycling of digester supernatant

14. The second-stage BOD is also referred to as the ______ stage. 14.____

A. carbonaceous B. pretreatment
C. flocculation D. nitrification

15. When organic matter decomposes to form foul-smelling products associated with the lack of free oxygen, this condition is known as 15.____

A. shock loading B. septicity
C. sloughing D. sidestreaming

16. Which type of bacteria has the HIGHEST optimum temperature for treatment? 16.____

A. Mesophilic B. Cryophilic
C. Thermophilic D. Psychrophilic

17. The COD test 17.____

A. estimates the total oxygen consumed
B. measures the carbon oxygen demand
C. provides results more quickly than the BOD test
D. measures only the nitrification oxygen demand

18. Which of the following is NOT considered a major factor that may cause variations in lab test results? 18.____

A. The nature of the material being examined
B. Testing equipment
C. Sampling procedures
D. The quantity of material being examined

19. The treatment process that MOST effectively removes suspended solids from wastewater is 19.____

A. sedimentation
B. flocculation
C. skimming
D. comminution

20. Which of the following is a thickening alternative in sludge processing? 20.____

A. Flotation
B. Incineration
C. Elutriation
D. Wet oxidation

21. The device that continuously adds the flow of wastewater into a plant is the 21.____

A. aggregate
B. turbidity meter
C. titrator
D. totalizer

22. Two types of measurement required in connection with the operation of a treatment plant are 22.____

A. effluent and downstream
B. temperature and dissolved oxygen
C. in-plant and receiving water
D. temperature and receiving water

23. You may NOT dispose of excess activated sludge waste from package plants 23.____

A. at a nearby treatment plant
B. by anaerobic digestion
C. by removal by septic tank pumper
D. by aeration in a holding tank, then deposit in a sanitary landfill

24. What is the term for the combination of activated sludge with raw wastewater in a treatment plant? 24.____

A. Median
B. Liquefaction
C. Effluent
D. Mixed liquor

25. Landfills produce poisonous ______ gas as a byproduct of decomposition. 25.____

A. methane
B. nitrogen
C. chlorofluorocarbons
D. argon

KEY (CORRECT ANSWERS)

1. B
2. D
3. A
4. D
5. B

6. B
7. C
8. D
9. D
10. B

11. A
12. C
13. A
14. D
15. B

16. C
17. C
18. D
19. B
20. A

21. D
22. C
23. B
24. D
25. A

TEST 2

DIRECTIONS: Each question or incomplete statement is followed by several suggested answers or completions. Select the one that BEST answers the question or completes the statement. *PRINT THE LETTER OF THE CORRECT ANSWER IN THE SPACE AT THE RIGHT.*

1. Which of the following types of pumps is a kinetic pump? 1.____

   A. Rotary B. Piston plunger
   C. Hydraulic ram D. Blow case

2. What device is used to keep floated solids out of the effluent in dissolved air flotation thickeners? 2.____

   A. Cloth screens B. Microscreens
   C. Effluent baffles D. Water sprays

3. The ______ is NOT one of the primary factors affecting the flow of wastewater and sewage in sewers. 3.____

   A. viscosity of the liquid
   B. cross-sectional area of the system conduit
   C. time of day
   D. pipe surface

4. What is the term for washing a digested sludge in the plant effluent? 4.____

   A. Masking B. Elutriation
   C. Hydrolysis D. Slaking

5. ______ is NOT an objective in periodically pumping sludge from the primary clarifier to the digester. 5.____

   A. Prevention of pump clogging
   B. Prevention of digester overload
   C. Allowance for thicker sludge pumping
   D. Maintenance of good clarifier conditions

6. The toxic chemical LEAST likely to be encountered by treatment plant operators is(are) 6.____

   A. mercury B. acids
   C. fluorocarbons D. bases

7. Which concentration of total dissolved solids, in milligrams per liter, would be the MINIMUM required in order to be considered *strong* in wastewater? 7.____

   A. 250 B. 500 C. 850 D. 1,200

8. What is the term for the treatment process in which a tank or reactor is filled, the water is treated, and the tank is emptied? 8.____

   A. Flocculation B. Centration
   C. Batch process D. Pond process

9. The mixing of a compound with water to produce a true chemical reaction is to 9.____

   A. dissolve B. slake C. strip D. hydrate

10. If the difference in elevation between inflow and outflow sewers is greater than 1.5 feet, which device is needed? 10.____

A. Side weir
B. Drop inlet
C. Baffles
D. Inlet casting

11. Intermittent releases or discharges of industrial wastes are known as 11.____

A. slurries
B. slugs
C. splashes
D. stop logs

12. Results from the settleability test of activated sludge solids may be used to 12.____

A. calculate BOD
B. determine probable flow rates at which sludges may clog equipment
C. calculate sludge age
D. determine ability of solids to separate from liquid in final clarifier

13. The device used to measure the temperature of an effluent is a 13.____

A. thermometer
B. Bourdon tube
C. thermocouple
D. pug mill

14. Which source is typically the HEAVIEST contributor of total solids in a service area's wastewater supply? 14.____

A. Industrial wastes
B. Domestic wash waters
C. Storm runoff
D. Human biological wastes

15. The term for liquid removed from a settled sludge is 15.____

A. hydrolyte
B. supernatant
C. aliquot
D. slurry

16. A unit of wastewater moving through the treatment system without dispersing or mixing with the rest of the wastewater in the system is called 16.____

A. centration
B. plug flow
C. putrefaction
D. slugging

17. What is the term for the groups or clumps of bacteria or particles that have clustered together during the treatment process? 17.____

A. Coagulants
B. Slurries
C. Floes
D. Slugs

18. The purpose of PRIMARY sedimentation is to remove 18.____

A. settleable and floatable material
B. roots, rags, and large debris
C. suspended and dissolved solids
D. sand and gravel

19. ______ would NOT cause an increase in effluent coliform levels at a treatment plant. 19.____

A. Mixing problems
B. An increase in effluent BOD
C. Solids accumulation in the contact chamber
D. High chlorine residual

20. What is the term used to describe bacteria that can live under either aerobic or anaerobic conditions? 20.____

A. Cultured
B. Agglomerated
C. Filamentous
D. Facultative

21. Which devices are NOT used during pretreatment? 21.____

A. Racks
B. Comminutors
C. Screens
D. Coagulators

22. Through which stage in an activated sludge treatment plant would wastewater pass FIRST? 22.____

A. Grit chambers
B. Bar racks
C. Settling tanks
D. Primary sedimentation

23. The inorganic gas LEAST likely to be found around a treatment plant is 23.____

A. ammonia
B. methane
C. hydrogen sulfide
D. mercaptans

24. The soils in an effluent disposal on land program may be tested using each of the following procedures EXCEPT 24.____

A. BOD
B. conductivity
C. pH
D. cation exchange capacity

25. Which of the following is a conditioning alternative in sludge processing? 25.____

A. Centrifugation
B. Drying
C. Composing
D. Elutriation

KEY (CORRECT ANSWERS)

1. C
2. C
3. C
4. B
5. A

6. C
7. C
8. C
9. B
10. B

11. B
12. D
13. C
14. A
15. B

16. B
17. C
18. A
19. D
20. D

21. D
22. B
23. D
24. A
25. D

READING COMPREHENSION
UNDERSTANDING AND INTERPRETING WRITTEN MATERIAL

EXAMINATION SECTION
TEST 1

DIRECTIONS: Each question or incomplete statement is followed by several suggested answers or completions. Select the one that BEST answers the question or completes the statement. *PRINT THE LETTER OF THE CORRECT ANSWER IN THE SPACE AT THE RIGHT.*

Questions 1-5.

DIRECTIONS: Questions 1 through 5 are to be answered SOLELY on the basis of the following paragraph.

The strength of the seal of a trap is closely proportional to the depth of the seal, regardless of the size of the trap. Unfortunately, an increase in the depth of the seal also increases the probability of solids being retained in the trap, and a limit of about a 4" depth of seal for traps that must pass solids has been imposed by some plumbing codes.The depth of seal most commonly found in simple traps is between $1\frac{1}{2}$" and 2". The Hoover Report recommends a minimum depth of 2" as a safeguard against seal rupture and a maximum depth of 4" to avoid clogging, fungus growths, and similar difficulties. Traps in rain-water leaders and other pipes carrying clear-water wastes only, and which are infrequently used, should have seal depths equal to or greater than 4". The increase in the volume of water retained in the trap helps very little in increasing the strength of the seal, but it does materially reduce the velocity of flow through the trap so as to increase the probability of the sedimentation of solids therein.

1. In accordance with the above, it may be said that traps carrying rain-water should have a seal of 1.____

   A. 5"
   B. $3\frac{1}{2}$"
   C. 2"
   D. $1\frac{1}{2}$"

2. In accordance with the preceding paragraph, which one of the following statements is MOST NEARLY correct? 2.____

   A. Simple traps have a depth of seal between $1\frac{1}{2}$" to 4".
   B. A minimum depth of 4" is recommended to avoid seal rupture.
   C. The strength of the seal is proportional to the size of the trap.
   D. The higher the depth of seal, the more chance of collecting solids.

3. In accordance with the above, it may be said that increasing the volume of water retained in a trap may 3.___

   A. *greatly* increase the velocity of flow
   B. *slightly* increase the velocity of flow
   C. *greatly* increase the trap seal
   D. *slightly* increase the trap seal

4. Of the following, the title which BEST explains the main idea of this paragraph is 4.___

   A. TRAP SEAL DEPTHS
   B. THE EFFECTS OF SEDIMENTATION ON TRAP SEALS
   C. COMMON TRAP SIZES
   D. TRAP SIZES AND VELOCITY OF FLOW

5. Assume that the strength of a trap seal is indicated by 8 units when the trap depth is 2". In accordance with the above paragraph, increasing the depth of seal to 4" will cause the strength of the trap seal to be MOST NEARLY ______ units. 5.___

   A. 2 B. 4 C. 8 D. 16

Questions 6-10.

DIRECTIONS: Questions 6 through 10 are to be answered SOLELY on the basis of the following paragraph.

The thickness of insulation necessary for the most economical results varies with the steam temperature. The standard covering consists of 85 percent magnesia with 10 percent of long-fibre asbestos as a binder. Both magnesia and laminated asbestos–felt and other forms of mineral wool including glass wool–are also used for heat insulation. The magnesia and laminated asbestos coverings may be safely used at temperatures up to 600° F. Pipe insulation is applied in molded sections 3 feet long. The sections are attached to the pipe by means of galvanized iron wire or netting. Flanges and fittings can be insulated by direct application of magnesia cement to the metal without reinforcement. Insulation should always be maintained in good condition because it saves fuel. Routine maintenance of warm-pipe insulation should include prompt repair of damaged surfaces. Steam and hot water leaks concealed by insulation will be difficult to detect. Underground steam or hotwater pipes are best insulated using a concrete trench with removable cover.

6. The word *reinforcement,* as used above, means MOST NEARLY 6.___

   A. resistance B. strengthening
   C. regulation D. removal

7. According to the above paragraph, magnesia and laminated-asbestos coverings may be safely used at temperatures up to 7.___

   A. 800° F B. 720° F C. 675° F D. 600° F

8. According to the above paragraph, insulation should ALWAYS be maintained in good condition because it 8.___

   A. is laminated B. saves fuel
   C. is attached to the pipe D. prevents leaks

9. According to the above paragraph, pipe insulation sections are attached to the pipe by means of 9.____

A. binders
B. mineral wool
C. netting
D. staples

10. According to the above paragraph, a leak in a hot-water pipe may be difficult to detect because when insulation is used, the leak is 10.____

A. underground
B. hidden
C. routine
D. cemented

Questions 11-15.

DIRECTIONS: Questions 11 through 15 are to be answered SOLELY on the basis of the following paragraph.

Reductions in pipe size of a building heating system are made with eccentric fittings and are pitched downward. The ends of mains with gravity return shall be at least 18" above the water line of the boiler. As condensate flows opposite to the stream, runouts are one size larger than the vertical pipe and are pitched upward. In a one-pipe system, an automatic air vent must be provided at each main to relieve air pressure and to let steam enter the radiator. As steam enters the radiator, a thermal device causes the vent to close, thereby holding the steam. Steam mains should not be less than two inches in diameter. The end of the steam main should have a minimum size of one-half of its greatest diameter. Small steam systems should be sized for a 2 oz. pressure drop. Large steam systems should be sized for a 4 oz. pressure drop.

11. The word *thermal*, as used in the above paragraph, means MOST NEARLY 11.____

A. convector B. heat C. instrument D. current

12. According to the above paragraph, the one of the following that is one size larger than the vertical pipe is the 12.____

A. steam main
B. valve
C. water line
D. runout

13. According to the above paragraph, small steam systems should be sized for a pressure drop of ______ ounces. 13.____

A. 2 B. 3 C. 4 D. 5

14. According to the above paragraph, ends of mains with gravity return shall be AT LEAST 14.____

A. 18" above the water line of the boiler
B. one-quarter of the greatest diameter of the main
C. twice the size of the vertical pipe in the main
D. 18" above the steam line of the boiler

15. According to the above paragraph, the one of the following that is provided at each main to relieve air pressure is a(n) 15.____

A. gravity return
B. convector
C. eccentric
D. vent

Questions 16-17.

DIRECTIONS: Questions 16 and 17 are to be answered SOLELY on the basis of the following paragraph.

In determining the size of a storm drain, a number of factors must be taken into consideration. One factor which makes sizing the storm drain difficult is the matter of predicting rainfall over a given period. Using a maximum estimate of about 1 inch of rain in a 10-minute interval, the approximate volume of water that will fall on a roof or surface in one minute's time can be determined readily. Another factor is the pitch and material of a roof or surface upon which the rain falls. A surface that has a pitch and smooth surface would increase the flow of water into a drain pipe.

16. According to the above paragraph, the statement which includes all factors needed to determine the size of a drain pipe is the 16.___

A. maximum rainfall on a surface
B. pitch and surface of the area
C. amount of water to be piped in a definite time interval
D. area of the surface

17. A roof that has a 45° pitch would PROBABLY have a drain pipe size 17.___

A. smaller than a roof with no pitch
B. larger than a roof with no pitch
C. equal to that of a flat roof
D. equal to the amount of water falling in ten minutes

Questions 18-19.

DIRECTIONS: Questions 18 and 19 are to be answered SOLELY on the basis of the following paragraph.

Because of the large capacity of unit heaters, care should be taken to see that the steam piping leading to them is of sufficient size. Unit heaters should not be used on one-pipe systems. If the heating system contains direct radiators operated with steam under vacuum, it is best to have the unit heaters served by a separate main so that steam above atmospheric pressure can be supplied to the units, if desired, without interfering with the operation of the direct radiators.

18. According to the above paragraph, unit heaters are supplied with 18.___

A. steam under vacuum
B. steam from direct radiators
C. separate steam lines
D. steam preferably from a one-pipe system

19. According to the above paragraph, it may be said that unit heaters work BEST with 19.___

A. steam above atmospheric pressure
B. direct radiators
C. one-pipe system
D. vacuum systems

Questions 20-21.

DIRECTIONS: Questions 20 and 21 are to be answered SOLELY on the basis of the following paragraph.

Most heating units emit heat by radiation and convection. An exposed radiator emits approximately half of its heat by radiation, the amount depending upon the size and number of sections. In general, a thin radiator, such as a wall radiator, emits a larger proportion of its heat by radiation than does a thick radiator. When a radiator is enclosed or shielded, the proportion of heat emitted by radiation is reduced. The balance of the emission occurs by conduction to the air in contact with the heating surface, and this heated air rises by circulation due to convection and transmits this warm air to the space which is to be heated.

20. According to the above paragraph, when a radiator is enclosed, a GREATER portion of the heat is emitted to the room by 20.____

A. convection
B. radiation
C. conduction
D. transmission

21. According to the above paragraph, the amount of heat that a radiator emits is 21.____

A. approximately half of its heat by radiation
B. determined by the thickness of the radiator
C. dependent upon whether it is exposed or enclosed
D. dependent upon the size and number of sections of the radiator

Questions 22-25.

DIRECTIONS: Questions 22 through 25 are to be answered SOLELY on the basis of the following paragraph.

Safety valves are required to operate without chattering and to be set to close after blowing down not more than 4% of the set pressure, but not less than 2 lbs. in any case. For pressure between 100 and 300 lbs., inclusive, the blow down is required to be not less than 2% of the set pressure. The blow down adjustment is made and sealed by the manufacturer. The popping-point tolerance plus or minus is required not to exceed 2 lbs. for pressure up to and including 70 lbs., 3 lbs. for pressure 71 to 300 lbs., and 10 lbs. for pressure over 300 lbs.

22. A boiler is being installed to operate at a maximum allowable pressure of 10 lb., and the safety valve has been set to blow at this pressure. 22.____
This valve should close after the boiler blows down to NOT MORE THAN ______ lb.

A. 9.6
B. 4.0
C. 9.8
D. 8.0

23. A boiler is being installed to operate at a maximum allowable working pressure of 300 lb., and the safety valve is set to blow at this pressure. This valve should close after the boiler blows down to NOT MORE THAN ______ lb. 23.____

A. 204
B. 298
C. 12
D. 6

24. A sealed safety valve is to be installed on a superheater header in a power steam generating plant. The marking on this valve shows that it is set to pop at 425 lb. 24.____
This valve would operate satisfactorily if it popped at EITHER ______ or ______ lb.

A. 425; 445
B. 415; 435
C. 372.5; 467.5
D. 412.25; 437.75

25. A sealed safety valve is to be installed on a boiler in a high pressure steam generating station. The marking on the valve shows that it is set to pop at 300 lb. 25.____
This valve would operate satisfactorily if it popped at EITHER ______ or ______ lb.

A. 290; 310
B. 297; 303
C. 291; 309
D. 288; 312

KEY (CORRECT ANSWERS)

1. A
2. D
3. D
4. A
5. D
6. B
7. D
8. B
9. C
10. B
11. B
12. D
13. A
14. A
15. D
16. C
17. B
18. C
19. A
20. A
21. D
22. D
23. A
24. B
25. B

TEST 2

DIRECTIONS: Each question or incomplete statement is followed by several suggested answers or completions. Select the one that BEST answers the question or completes the statement. *PRINT THE LETTER OF THE CORRECT ANSWER IN THE SPACE AT THE RIGHT.*

Questions 1-6.

DIRECTIONS: Questions 1 through 6 are to be answered SOLELY on the basis of the following paragraph.

FIRST AID INSTRUCTIONS

The main purpose of first aid is to put the injured person in the best possible position until medical help arrives. This includes the performance of emergency treatment for the purpose of saving a life if a doctor is not present. When a person is hurt, a crowd usually gathers around the victim. If nobody uses his head, the injured person fails to get the care he needs. You must stay calm and, most important, it is your duty to take charge at an accident. The first thing for you to do is to see, as best you can, what is wrong with the injured person. Leave the victim where he is until the nature and extent of his injury are determined. If he is unconscious, he should not be moved, except to lay him flat on his back if he is in some other position. Loosen the clothing of any seriously hurt person, and make him as comfortable as possible. Medical help should be called as soon as possible. You should remain with the injured person and send someone else to call the doctor. You should try to make sure that the one who calls for a doctor is able to give correct information as to the location of the injured person. In order to help the physician to know what equipment may be needed in each particular case, the person making the call should give the doctor as much information about the injury as possible.

1. If nobody uses his head at the scene of an accident, there is danger that 1.____

   A. no one will get the names of all the witnesses
   B. a large crowd will gather
   C. the victim will not get the care he needs
   D. the victim will blame the city for negligence

2. When an accident occurs, the FIRST thing you should do is 2.____

   A. call a doctor
   B. loosen the clothing of the injured person
   C. notify the victim's family
   D. try to find out what is wrong with the injured person

3. If you do NOT know the extent and nature of the victim's injuries, you should 3.____

   A. let the injured person lie where he is
   B. immediately take the victim to a hospital yourself
   C. help the injured person to his feet to see if he can walk
   D. have the injured person sit up on the ground while you examine him

4. If the injured person is breathing and unconscious, you should 4.___

   A. get some hot liquid such as coffee or tea into him
   B. give him artificial respiration
   C. lift up his head to try to stimulate blood circulation
   D. see that he lies flat on his back

5. If it is necessary to call a doctor, you should 5.___

   A. go and make the call yourself since you have all the information
   B. find out who the victim's family doctor is before making the call
   C. have someone else make the call who know the location of the victim
   D. find out which doctor the victim can afford

6. It is important for the caller to give the doctor as much information as is available regarding the injury so that the doctor 6.___

   A. can bring the necessary equipment
   B. can make out an accident report
   C. will be responsible for any malpractice resulting from the first aid treatment
   D. can inform his nurse on how long he will be in the field

Questions 7-8.

DIRECTIONS: Questions 7 and 8 are to be answered SOLELY on the basis of the following paragraph.

PRECIPITATION AND RUNOFF

In the United States, the average annual precipitation is about 30 inches, of which about 21 inches is lost to the atmosphere by evaporation and transpiration. The remaining 9 inches becomes runoff into rivers and lakes. Both the precipitation and runoff vary greatly with geography and season. Annual precipitation varies from more than 100 inches in parts of the northwest to only 2 or 3 inches in parts of the southwest. In the northeastern part of the country, including New York State, the annual average precipitation is about 45 inches, of which about 22 inches becomes runoff. Even in New York State, there is some variation from place to place and considerable variation from time to time. During extremely dry years, the precipitation may be as low as 30 inches and the runoff below 10 inches. In general, there are greater variations in runoff rates from smaller watersheds. A critical water supply situation occurs when there are three or four abnormally dry years in succession.

Precipitation over the state is measured and recorded by a network of stations operated by the U.S. Weather Bureau. All of the precipitation records and other data such as temperature, humidity, and evaporation rates are published monthly by the Weather Bureau in *Climatological Data.* Runoff rates at more than 200 stream-gauging stations in the state are measured and recorded by the U.S. Geological Survey in cooperation with various state agencies. Records of the daily average flows are published annually by the U.S. Geological Survey in *Surface Water Records of New York.* Copies may be obtained by writing to the Water Resources Division, United States Geological Survey, Albany, New York 23301.

7. From the above paragraphs, it is APPROPRIATE to conclude that 7.____

   A. critical supply situations do not occur
   B. the greater the rainfall, the greater the runoff
   C. there are greater variations in runoff from larger watersheds
   D. the rainfall in the southwest is greater than the average in the country

8. From the above paragraphs, it is APPROPRIATE to conclude that 8.____

   A. an annual rainfall of about 50 inches does not occur in New York State
   B. the U.S. Weather Bureau is only interested in rainfall
   C. runoff is equal to rainfall less losses to the atmosphere
   D. information about rainfall and runoff in New York State is unavailable to the public

Questions 9-10.

DIRECTIONS: Questions 9 through 10 are to be answered SOLELY on the basis of the following paragraph.

NATURAL LAKES

Large lakes may yield water of exceptionally fine quality except near the shore line and in the vicinity of sewer outlets or near outlets of large streams. Therefore, minimum treatment is required. The availability of practically unlimited quantities of water is also a decided advantage. Unfortunately, however, the sewage from a city is often discharged into the same lake from which the water supply is taken. Great care must be taken in locating both the water intake and the sewer outlet so that the pollution handled by the water treatment plant is a minimum.

Sometimes the distance from the shore where dependable, satisfactory water can be found is so great that the cost of water intake facilities is prohibitive for a small municipality. In such cases, another supply must be found, or water must be obtained from a neigh-boring large city. Lake water is usually uniform in quality from day to day and does not vary in temperature as much as water from a river or small impounding reservoir.

9. A DISADVANTAGE of drawing a water supply from a large lake is that 9.____

   A. expensive treatment is required
   B. a limited quantity of water is available
   C. nearby cities may dump sewage into the lake
   D. the water is too cold

10. An ADVANTAGE of drawing a water supply from a large lake is that the 10.____

   A. water is uniform in quality
   B. water varies in temperature
   C. intake is distant from the shore
   D. intake may be near a sewer outlet

Questions 11-13.

DIRECTIONS: Questions 11 through 13 are to be answered SOLELY on the basis of the following paragraph.

Excavation of trench–The trench shall be excavated as directed; one side of the street or avenue shall be left open for traffic at all times. In paved streets, the length of trench that may be opened between the point where the backfilling has been completed and the point where the pavement is being removed shall not exceed fifteen hundred feet for pipes 24 inches or less in diameter. For pipes larger than 24-inch, the length of open trenches shall not exceed one thousand feet. The completion of the backfilling shall be interpreted to mean the backfilling of the trench and the consolidation of the backfill so that vehicular traffic can be resumed over the backfill, and also the placing of any temporary pavement that may be required.

11. According to the above paragraph, the street 11.___

A. can be closed to traffic in emergencies
B. can be closed to traffic only when laying more than 1500 feet of pipe
C. is closed to traffic as directed
D. shall be left open for traffic at all times

12. According to the above paragraph, the MAXIMUM length of open trench permitted in paved streets depends on the 12.___

A. traffic on the street
B. type of ground that is being excavated
C. water conditions met with in excavation
D. diameter of the pipe being laid

13. According to the above paragraph, the one of the following items that is included in the *completion of the backfilling* is 13.___

A. sheeting and bracing
B. cradle
C. temporary pavement
D. bridging

Questions 14-16.

DIRECTIONS: Questions 14 through 16 are to be answered SOLELY on the basis of the following paragraph.

The Contractor shall notify the Engineer by noon of the day immediately preceding the date when he wishes to shut down any main; and if the time set be approved, the Contractor shall provide the men necessary to shut down the main at the time stipulated, and to previously notify all consumers whose supply may be affected. These men shall be under the direction of the Department employees, who will superintend all operations of valves and hydrants. Shutdowns for making connections will not be made unless and until the Contractor has everything on the ground in readiness for the work.

14. According to the above paragraph, before a contractor can make a shut-down, he MUST notify the 14.___

A. police department
B. District Foreman
C. Engineer
D. Highway Department

15. According to the above paragraph, the operation of the valves will be supervised by the 15.____

A. Department employees
B. Contractor's men
C. Contractor's superintendent
D. Engineer

16. According to the above paragraph, shut-downs for connections are made 16.____

A. the day before the connection is to be made
B. first and then consumers are notified
C. at any time convenient to the Contractor
D. when the Contractor has everything on the ground in readiness for the work

Questions 17-22.

DIRECTIONS: Questions 17 through 22 are to be answered SOLELY on the basis of the following paragraphs.

HOT WATER GENERATION

The hot water that comes from a faucet is called Domestic Hot Water. It is heated by a steam coil that runs through a storage tank full of water in the basement of each building.

As the tenants take the hot water, fresh cold water enters the tank and is heated. The temperature of this water is automatically kept at approximately 140° F.

The device which controls the temperature is called a temperature regulator valve. It is operated by a bellows, capillary tube, and thermo bulb which connects between the valve and the hot water being stored in the tank. This bulb, tube, and bellows contains a liquid which expands and contracts with changes in temperatures.

As the water in the tank reaches 140° F, the liquid in the thermo bulb expands and causes pressure to travel along the capillary tube and into the bellows. The expanded liquid forces the bellows to push the Temperature Regulator Valve Stem down, closing the valve. No more steam can enter the coil in the tank, and the water will get no hotter.

As the hot water is used by the tenants, cold water enters the tank and pulls the temperature down. This causes the liquid in the thermo bulb to cool and contract (shrink). The pressure is no longer in the bellows, and a spring pushes it up, allowing the valve to open and allowing steam to again enter the heating coil in the storage tank, raising the temperature of the Domestic Hot Water to 140° F.

17. Domestic hot water is heated by 17.____

A. coal
B. electricity
C. hot water
D. steam

18. The temperature of domestic hot water is MOST NEARLY 18.____

A. 75° F
B. 100° F
C. 140° F
D. 212° F

19. The temperature of the hot water is controlled by a 19.___

A. thermometer
B. temperature regulator valve
C. pressuretrol
D. pressure gauge

20. The temperature regulator valve is operated by a combination of a 20.___

A. thermometer and a thermo bulb
B. thermometer and a pyrometer
C. bellows, capillary tube, and a thermometer
D. bellows, capillary tube, and a thermo bulb

21. Closing of the temperature regulator valve prevents ______ from entering the heating coil in the tank. 21.___

A. water
B. steam
C. electricity
D. air

22. As hot water is used by the tenants, the temperature of the water in the tank 22.___

A. increases
B. decreases
C. remains the same
D. approaches 212° F

Question 23.

DIRECTIONS: Question 23 is to be answered SOLELY on the basis of the following paragraph.

Lack of service meters has a definite effect on water consumption. Metering of all services of a city should reduce consumption to about 50 percent of the consumption without meters. Although metering reduces water consumption, there is a tendency for consumption to increase gradually after all services are metered.

23. According to the above paragraph, the one of the following statements that is CORRECT is: 23.___

A. Consumption of water is cut approximately in half by metering, but once all services are metered, the consumption then increases gradually
B. After all services are metered, water consumption continues to decrease steadily
C. Metering of all services reduces the consumption of water by much more than half
D. Water consumption is not affected by metering of all services

Question 24.

DIRECTIONS: Questions 24 is to be answered SOLELY on the basis of the following paragraph.

A venturi meter operates without moving parts and hence is the simplest type of meter in use so far as its construction is concerned. It is a velocity meter, and it is suitable for measuring only high rates of flow. Rates of flow below its capacity limit are not accurately measured. It is, therefore, not suitable for use in measuring the low intermittent demand of most consumers.

24. According to the above paragraph, the flow in a pipe which would MOST accurately be measured by a venturi meter is 24.____

A. an intermittent flow below the meter's capacity
B. a steady flow below the meter's capacity
C. a steady flow at the meter's capacity
D. intermittent flows above or below capacity of the meter

Question 25.

DIRECTIONS: Question 25 is to be answered SOLELY on the basis of the following paragraph.

A house service water supply connection may be taken from the sprinkler water supply connection to the public main if the diameter of the house service water supply connection is not greater than onehalf the diameter of the sprinkler water supply connection. No shutoff valve shall be placed on the sprinkler supply line other than the main shut-off valve for the building on the street side of the house service water supply connection. If such a connection is made and if a tap also exists for the house service water supply, the tap shall be plugged.

25. According to the above paragraph, the one of the following statements that is CORRECT is: 25.____

A. A sprinkler water supply connection should be at least twice the diameter of any house service water supply connection taken from it
B. A shut-off valve, in addition to the main shut-off valve, is required on sprinkler supply lines on the street side of the house service water supply connection
C. Where a house service water supply is connected to the sprinkler water supply and there is a tap for the house service water supply, the tap may remain in service
D. A house service water supply connection may be taken off each side of the main shut-off valve of the sprinkler water supply

---

KEY (CORRECT ANSWERS)

1. C
2. D
3. A
4. D
5. C
6. A
7. B
8. C
9. C
10. A
11. D
12. D
13. C
14. C
15. A
16. D
17. D
18. C
19. B
20. D
21. B
22. B
23. A
24. C
25. A

TEST 3

DIRECTIONS: Each question or incomplete statement is followed by several suggested answers or completions. Select the one that BEST answers the question or completes the statement. *PRINT THE LETTER OF THE CORRECT ANSWER IN THE SPACE AT THE RIGHT.*

Questions 1-4.

DIRECTIONS: Questions 1 through 4 are to be answered SOLELY on the basis of the following paragraph.

Welds in sheet metal up to 1/16 inch in thickness can be made satisfactorily by flanging the edges of the joint. The edges are prepared by turning up a very thin lip or flange along the line of the joint. The height of this flange should be equal to the thickness of the sheet being welded. The edges should be aligned so that the flanges stand up, and the joint should be tack-welded every 5 or 6 inches. Heavy angles or bars should be clamped on each side of the joint to prevent distortion or buckling. No filler metal is required for making this joint. The raised edges are quickly melted by the heat of the welding flame so as to produce an even weld bead which is nearly flush with the original sheet metal surface. By controlling the speed of welding and the motion of the flame, good fusion to the under side of the sheets can be obtained without burning through.

1. According to the above paragraph, satisfactory welds may be made in sheet metal by flanging the edges. 1.___
The MAXIMUM thickness of metal recommended is

A. 20 gauge
B. 18 gauge
C. 1/16"
D. 5/64"

2. According to the above paragraph, good fusion may be obtained without burning through of the metal by controlling the motion of the flame and the 2.___

A. size of tip
B. speed of welding
C. oxygen flow
D. acetylene flow

3. According to the above paragraph, if the thickness of the metal is 1/32", then the flange height should be 3.___

A. 1/64"
B. 1/32"
C. 1/16"
D. 1/8"

4. According to the above paragraph, distortion in the welding of sheet metal may be prevented by 4.___

A. controlling the speed of welding
B. use of a flange of correct height
C. use of proper filler metal
D. clamping angles on each side of the joint

Questions 5-12.

DIRECTIONS: Questions 5 through 12 are to be answered SOLELY on the basis of the Edison storage battery maintenance procedure below.

EDISON STORAGE BATTERY MAINTENANCE PROCEDURE

Take a voltage reading of each cell in the battery with a voltmeter. Any battery with two or more dead or reverse cells is to be removed and sent to the shop. All cell caps are to be opened, and the water level brought up to 2 3/4" above the plates. Any battery requiring a considerable amount of water must be called to the foreman's attention. All cell caps must be brushed clean and Edison battery oil applied to them. No batteries are to remain in service with cell caps broken or missing. The normal specific gravity reading of the solution must not be above 1.230 nor below 1.160. This reading is to be taken only on batteries which are found to be weak. Batteries with specific gravity lower than 1.160 must be sent to the shop. Be careful when disconnecting leads from the battery since a slight, turn of the connecting post will result in a dead cell due to the cell plates becoming short-circuited. When disconnecting leads, use a standard Edison terminal puller. When recording defective cells, give the battery number, the car number, and the position of the cell in the battery. No. 1 cell is the cell to which the positive battery lead is connected and so on up to the last cell, No. 26, to which the negative lead is connected.

5. A normal specific gravity reading would be 5.____

A. 1.450 B. 1.294 C. 1.200 D. 1.180

6. Batteries with below normal specific gravity reading MUST 6.____

A. always have water added
B. be called to the foreman's attention
C. not be given a voltmeter test
D. be sent to the shop

7. The battery leads are disconnected by using 7.____

A. gas pliers
B. Edison battery oil to free them
C. a screwdriver to pry them off
D. a standard Edison terminal puller

8. To completely record a defective cell, ______ required. 8.____

A. only one identifying number is
B. two identifying numbers are
C. three identifying numbers are
D. four identifying numbers are

9. A battery MUST be taken out of service if it has 9.____

A. one dead cell
B. broken cell caps
C. one reversed cell
D. a low water level

10. The battery water level should be brought up above the plates by ______ inches. 10.____

A. 2.75 B. 1.370 C. 1.264 D. 0.600

11. Specific gravity readings are to be taken only on batteries which 11._

A. are removed from service
B. have missing cell caps
C. are weak
D. have a high water level

12. Dead cells are sometimes caused by 12._

A. a slight turn of the connecting post
B. taking unnecessary gravity readings
C. adding too little battery oil
D. adding too much water

Questions 13-14.

DIRECTIONS: Questions 13 and 14 are to be answered SOLELY on the basis of the following paragraph.

It cannot be stressed too strongly that the greatest care should be taken in handling tools. If they are handled carelessly, serious accidents may result. Many accidents can be avoided if the back of the trowel is kept clean and if the trowel is not allowed to contain too much mortar. Where there is an *excess* of mortar, some might drop or splash into the plasterer's eyes. Any mortar which is dropped onto the hands, wrists, ankles, or underclothing should be removed immediately.

13. The MAIN point of the above paragraph is that 13._

A. all accidents will be avoided if tools are kept clean
B. most accidents can be avoided by the use of protective gloves
C. many accidents are caused by careless handling of tools
D. trowels should be kept free of mortar at all times

14. In the above paragraph, the word *excess* means MOST NEARLY 14._

A. surplus B. minor C. scant D. short

Questions 15-18.

DIRECTIONS: Questions 15 through 18 are to be answered SOLELY on the basis of the following paragraph.

There are two unfounded ideas that must be discarded before tackling the lube-simplification job. *Oil is oil* was a common expression from the middle of the nineteenth century up to the early 1900s. Then, as the century got well underway, *the pendulum swung in a wide arc.* At present, we find many oils being used, each with supposedly special properties. The large number of lube oils used at present results from the rapid growth at the same time of machine development and oil refining. The refiner acts to market new oils for each machine developed, and the machine manufacturer feels that each new mechanical unit is different from the others and needs a special lube oil. These feelings may be well-founded, but in many cases they are based on misinformation or blind faith in certain lube oil qualities. At the present time, operators and even lube engineers are finding it tough to keep track of all the claimed properties of all the lube oils.

15. It follows from the sense of this paragraph that the idea that *oil is oil* is unfounded because 15.____

A. it was conceived in the middle of the nineteenth century
B. the basic and varying properties of lube oils have now been shown to exist
C. lube oil properties, though fully known, were kept secret for economic reasons
D. there was no need for but one basic lube oil in the latter part of the nineteenth century

16. In the above paragraph, the phrase *the pendulum swung in a wide arc* means MOST NEARLY 16.____

A. oil refining was unable to keep up with machinery development
B. before 1900, lube oil engineers found it difficult to keep track of lube oil characteristics
C. the simplification of lube oils and their application was developed about 1900
D. many different lube oils with varying characteristics were marketed

17. As indicated in this paragraph, the simplification of the characteristics and the uses of lube oils is needed because the 17.____

A. manufacturers develop new machines to overcome competition
B. change in process at the refineries for a new lube oil is costly
C. present market is flooded with many so-called *special purpose* lube oils
D. *blind faith* of the operators in lube oil qualities should be rewarded

18. A reason given for the claimed need for special lube oil, as indicated in this paragraph, is that 18.____

A. development of new lube oils created the need for new machine units
B. lube oil engineers developed new tests and standards
C. basic crudes, from which lube oil is obtained, allow different refining methods
D. newly developed machines are so very different from each other

Questions 19-22.

DIRECTIONS: Questions 19 through 22 are to be answered SOLELY on the basis of the following paragraph.

ACCIDENT PREVENTION

Many accidents and injuries can be prevented if employees learn to be more careful. The wearing of shoes with thin or badly worn soles or open toes can easily lead to foot injuries from tacks, nails, and chair and desk legs. Loose or torn clothing should not be worn near moving machinery. This is especially true of neckties, which can very easily become caught in the machine. You should not place objects so that they block or partly block hallways, corridors, or other passageways. Even when they are stored in the proper place, tools, supplies, and equipment should be carefully placed or piled so as not to fall, nor have anything stick out from a pile. Before cabinets, lockers, or ladders are moved, the tops should be cleared of anything which might injure someone or fall off. If necessary, use a dolly to move these or other bulky objects.

Despite all efforts to avoid accidents and injuries, however, some will happen. If an employee is injured, no matter how small the injury, he should report it to his supervisor and have the injury treated. A small cut that is not attended to can easily become infected and can cause more trouble than some injuries which at first seem more serious. It never pays to take chances.

19. According to the above passage, the one statement that is NOT true is that 19.___

A. by being more careful, employees can reduce the number of accidents that happen
B. women should wear shoes with open toes for comfort when working
C. supplies should be piled so that nothing is sticking out from the pile
D. if an employee sprains his wrist at work, he should tell his supervisor about it

20. According to the above passage, you should NOT wear loose clothing when you are 20.___

A. in a corridor
B. storing tools
C. opening cabinets
D. near moving machinery

21. According to the above passage, before moving a ladder, you should 21.___

A. test all rungs
B. get a dolly to carry the ladder at all times
C. remove everything from the top of the ladder which might fall off
D. remove your necktie

22. According to the above passage, an employee who gets a slight cut should 22.___

A. have it treated to help prevent infection
B. know that a slight cut becomes more easily infected than a big cut
C. pay no attention to it as it can't become serious
D. realize that it is more serious than any other type of injury

Questions 23-25.

DIRECTIONS: Questions 23 through 25 are to be answered SOLELY on the basis of the following paragraph.

Keeping the city operating day and night requires the services of more than 400,000 civil service workers – roughly the number of people who live in Syracuse. This huge army of specialists works at more than 2,000 different jobs. The city's civil service workers are able to do everything that needs doing to keep our city running. Their only purpose is the well-being, comfort, and safety of the citizens of the city.

23. Of the following titles, the one that MOST nearly gives the meaning of the above paragraph is 23.___

A. CIVIL SERVICE IN SYRACUSE
B. EVERYONE WORKS
C. JOB VARIETY
D. SERVING THE CITY

24. According to the above paragraph, in order to keep the city operating 24 hours a day, 24.____

A. half of the civil service workers work days and half work nights
B. more than 400,000 civil service workers are needed on the day shift
C. the city needs about as many civil service workers as there are people in Syracuse
D. the services of some people who live in Syracuse is required

25. According to the above paragraph, it is MOST reasonable to assume that in the city's civil service, 25.____

A. a worker can do any job that needs doing
B. each worker works at a different job
C. some workers work at more than one job
D. some workers work at the same jobs

KEY (CORRECT ANSWERS)

1. C
2. B
3. B
4. D
5. C
6. D
7. D
8. C
9. B
10. A
11. C
12. A
13. C
14. A
15. B
16. D
17. C
18. D
19. B
20. D
21. C
22. A
23. D
24. C
25. D

MECHANICAL APTITUDE
TOOL RECOGNITION AND USE

EXAMINATION SECTION
TEST 1

DIRECTIONS: Each question or incomplete statement is followed by several suggested answers or completions. Select the one that BEST answers the question or completes the statement. *PRINT THE LETTER OF THE CORRECT ANSWER IS THE SPACE AT THE RIGHT.*

Questions 1-16.

DIRECTIONS: Questions 1 through 16 refer to the tools shown below. The numbers in the answers refer to the numbers below the tools.
NOTE: These tools are NOT shown to scale.

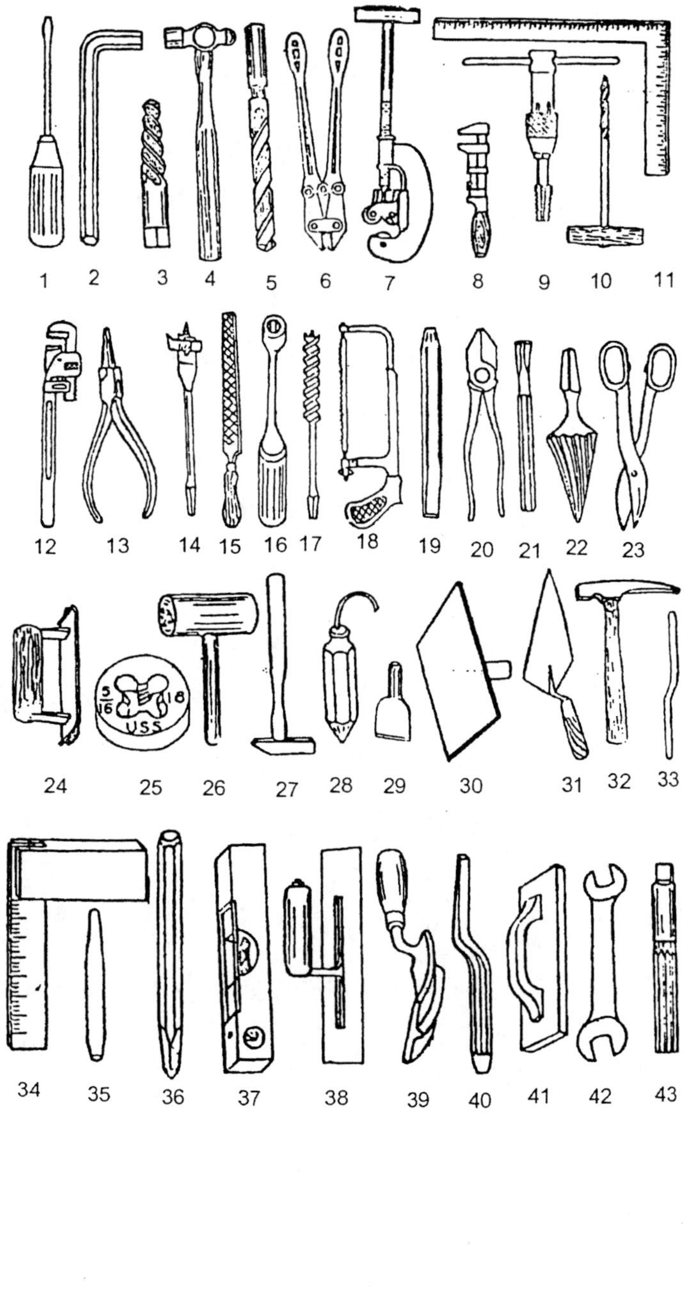

1. In order to cut a piece of 5/16" diameter steel scaffold hoisting cable, you should use tool number 1.____

A. 6 B. 7 C. 19 D. 23

2. Scaffold planks are secured to joisting irons by means of lag screws. To *properly* tighten these lag screws,you should use tool number 2.____

A. 12 B. 13 C. 20 D. 42

3. While installing a steel angle iron lintel,you find that the threads on the embedded holding bolts are damaged. You should repair the threads by using tool number 3.____

A. 7 B. 9 C. 25 D. 43

4. It is necessary to cut a hole in a concrete foundation wall in order to place a small bolt. To cut this small hole, you should use tool number 4.____

A. 14 B. 19 C. 21 D. 40

5. If tool number 17 bears the mark "7," this tool should be used to drill holes having a diameter of 5.____

A. 7/64" B. 7/32" C. 7/16" D. 7/8"

6. If the marking on the blade of tool number 18 reads "10-18," the "18" refers to the 6.____

A. number of teeth per inch
B. weight
C. thickness
D. length

7. If two points are separated by a vertical distance of 12 feet, the tool that should be used to make certain that the points are in *perfect* vertical alignment is number 7.____

A. 11 B. 28 C. 34 D. 37

8. A 3/4" diameter hole must be made in a steel floor beam. The tool you should use is number 8.____

A. 3 B. 5 C. 9 D. 22

9. To cut the corner off a building brick, you should use tool number 9.____

A. 4 B. 27 C. 29 D. 36

10. A 2"x2"x3/16" steel angle should be cut using tool number 10.____

A. 6 B. 7 C. 18 D. 19

11. The term "snips" should be applied to tool number 11.____

A. 6 B. 13 C. 20 D. 23

12. To line-up the bolt holes in two structural steel beams, you should use tool number 12.____

A. 1 B. 33 C. 35 D. 36

13. A "hawk" is tool number 13.____

A. 29 B. 30 C. 38 D. 41

14. After an 8" thick brick wall has been erected, it is discovered that a hole should have been left for a 4" sewer pipe. To cut that hole, you should use tool number 14.____

A. 5 B. 19 C. 32 D. 36

15. A "float" is tool number 15.____

A. 30 B. 31 C. 33 D. 41

16. A "Stillson" is tool number 16.____

A. 2 B. 8 C. 12 D. 22

KEY (CORRECT ANSWERS)

1. A
2. D
3. C
4. C
5. C
6. A
7. B
8. B
9. C
10. C
11. D
12. C
13. B
14. D
15. D
16. C

TEST 2

DIRECTIONS: Each question or incomplete statement is followed by several suggested answers or completions. Select the one that BEST answers the question or completes the statement. *PRINT THE LETTER OF THE CORRECT ANSWER IS THE SPACE AT THE RIGHT.*

1. The stake shown in the sketch at the right is a _______ stake. 1.____
   A. hatchet
   B. conductor
   C. solid mandrel
   D. beak horn

2. When a circle is too large to be drawn with a pair of dividers, the *proper* tool to use is a 2.____
   A. trammel
   B. protractor
   C. combination set
   D. flexible curve

3. A rivet set is a tool used to 3.____
   A. shape the head of a rivet
   B. mark off the spacing of rive
   C. remove a loose rivet
   D. check the shank length of a rivet

4. The hammer shown in the sketch at the right is a _______ hammer. 4.____
   A. raising
   B. ball peen
   C. setting
   D. cross-over

5. Of the following, the BEST tool to use to scribe a line parallel to the straight edge of a piece of sheet metal is a(n) 5.____
   A. outside caliper
   B. pair of dividers
   C. template
   D. scratch gage

6. Of the following, the BEST device to use to check the condition of the insulation of a cable is the 6.____
   A. ohmmeter
   B. wheatstone bridge
   C. voltmeter
   D. megger

7. Of the following fittings, the *one* used to connect two lengths of conduit in a straight line is a(n) 7.____
   A. elbow
   B. nipple
   C. tee
   D. coupling

8. If a nut is to be tightened to an exact specified value, the wrench that should be used is a(n) _______ wrench. 8._____

A. torque B. lock-jaw C. alligator D. spaner

9. A stillson wrench is *also* called a _______ wrench. 9._____

A. strap B. pipe C. monkey D. crescent

10. A machine screw is indicated on a drawing as  10._____
The head is the American Standard type called
A. flat
B. oval
C. fillister
D. round

11. The tool that is shown at the right is *properly* referred to as a (n) _______ tap. 11._____

A. bottoming
B. acme
C. taper
D. plug

12. The tool indicated at the right is referred to as an Arch Punch. 12._____
This tool should be used to

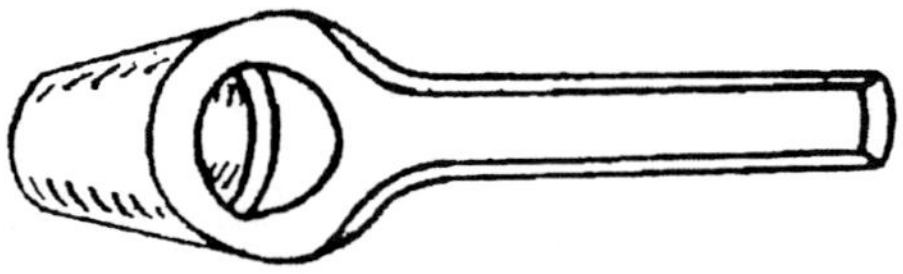

A. cut holes in 1/16" steel
B. cut large diameter holes in masonry
C. run through a conduit prior to pulling a cable or wires
D. make holes in rubber or leather gasket material

13. The plumbing fitting shown at the right is called a 13._____

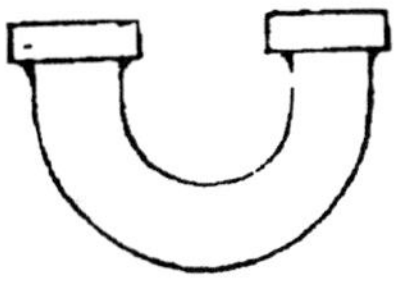

A. street elbow
B. return bend
C. running trap
D. reversing "el"

14. For which one of the following uses would it be *unsafe* to use a carpenter's hammer? Striking a 14._____

A. casing mail B. hand punch
C. hardened steel surface D. plastic surface

15. Of the following, the MAIN advantage in using a Phillips head screw is that 15._____

A. the threads of the Phillips head screw have a deeper bite than standard screw threads
B. the screwdriver used on this type of screw is more likely to keep its edge than a standard screwdriver
C. a single screwdriver fits all size screws of this type
D. the screwdriver used on this type of screw is less likely to slip than a standard screwdriver

16. *One* of the reasons why a polyester rope is considered to be the BEST general purpose rope is that it 16.____

A. does not stretch as much as ropes made of other materials
B. is available in longer lengths than ropes made of other materials
C. does not fray as much as ropes made of other materials
D. contains more strands than ropes made of other materials

17. The *proper* saw to use to cut wood with the grain is a _______ saw , 17.____

A. hack B. crosscut C. back D. rip

18. Assume that the instruction manual for a machine indicates that a certain bolt must be tightened with a specified amount of force. Of the following tools, the *one* which should be used to tighten the bolt with the specified amount of force is a(n) _______ wrench. 18.____

A. torque B. adjustable C. stillson D. combination

19. The power source of a pneumatic tool is 19.____

A. manuol
B. water pressure
C. compressed air
D. electricity

20. The tool used to cut internal pipe threads is a 20.____

A. broach B. tap C. die D. rod

KEY (CORRECT ANSWERS)

1.	A	11.	A
2.	A	12.	D
3.	A	13.	B
4.	C	14.	C
5.	D	15.	D
6.	D	16.	A
7.	D	17.	D
8.	A	18.	A
9.	B	19.	C
10.	B	20.	B

ARITHMETICAL REASONING
EXAMINATION SECTION
TEST 1

DIRECTIONS: Each question or incomplete statement is followed by several suggested answers or completions. Select the one that BEST answers the question or completes the statement. *PRINT THE LETTER OF THE CORRECT ANSWER IN THE SPACE AT THE RIGHT.*

1. 1.____

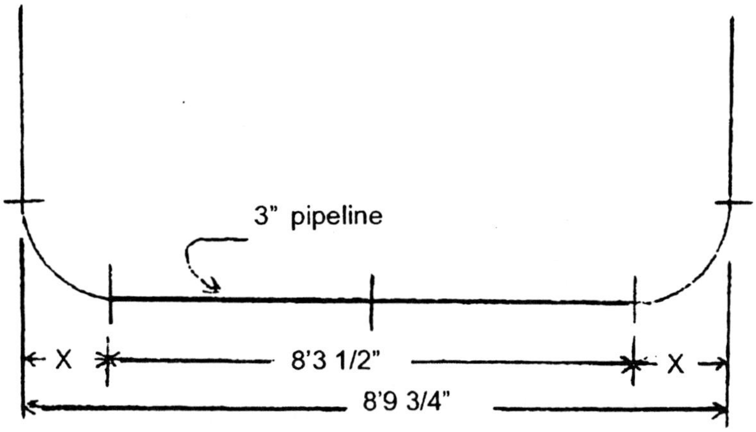

In the above sketch of a 3" pipeline, the distance X is MOST NEARLY ______ inches.

A. 3 1/8 B. 3 1/2 C. 3 1/2 D. 3 5/8

2. The fraction 9/64 is MOST NEARLY equal to 2.____

A. .1375 B. .1406 C. .1462 D. .1489

3. The sum of the following dimensions 1'2 3/16",1'5 1/2", and 1'4 5/8" is 3.____

A. 3'11 15/16" B. 4' 5/16"
C. 4'11/16" D. 4'1 5/8"

4. The scale on a plumbing drawing is 1/8" = 1 foot. 4.____
A horizontal line measuring 3 5/16" on the drawing would represent a length of ______ feet.

A. 24.9 B. 26.5 C. 28.3 D. 30.2

5. Assume that a water meter reads 50,631 cubic feet and the previous reading was 39,842 cubic feet. 5.____
If the charge for water is 23¢ per 100 cubic feet or any fraction thereof, the bill for the amount of water used since the previous meter reading will be

A. $24.22 B. $24.38 C. $24.84 D. $24.95

6. At a certain premises, the water consumption was 4 percent higher in 2015 than it was in 2014. 6.____
If the water consumption for 2015 was 9,740 cubic feet, then the water consumption for 2014 was MOST NEARLY ______ cubic feet.

A. 9,320 B. 9,350 C. 9,365 D. 9,390

7. A pump delivers water at a constant rate of 40 gallons per minute. 7.____
If there are 7.5 gallons to a cubic foot of water, the time it will take to fill a tank 6 feet x 5 feet x 4 feet is MOST NEARLY ______ minutes.

A. 15 B. 22.5 C. 28.5 D. 30

8. The total weight, in pounds, of three lengths of 3" cast-iron pipe 7'6" long, weighing 14.5 pounds per foot, and four lengths of 4" cast-iron pipe each 5'0" long, weighing 13.0 pounds per foot, is MOST NEARLY 8.____

A. 540 B. 585 C. 600 D. 665

9. The water pressure at the bottom of a column of water 34 feet high is 14.7 lbs./sq.in. 9.____
The water pressure in lbs./sq.in. at the bottom of the column of water 12 feet high is MOST NEARLY

A. 3 B. 5 C. 7 D. 9

10. The number of cubic yards of earth that would be removed when digging a trench 8 feet wide x 9 feet deep x 63 feet long is 10.____

A. 56 B. 168 C. 314 D. 504

11. On test, a meter registered one cubic foot for each 1 1/3 cubic feet of water that passed through it. 11.____
If the meter had a reading of 1,200 cubic feet, we may conclude that the CORRECT amount should be ______ cubic feet.

A. 800 B. 900 C. 1,500 D. 1,600

12. A water use meter reads 87,463 cubic feet. 12.____
If the previous reading was 17,377 cubic feet and the rate charged is 15 cents per 100 cubic feet, the bill for water use during this period is about

A. $45.00 B. $65.00 C. $85.00 D. $105.00

13. Under proper conditions, the one of the following groups of pipes that gives the same flow in gals/min as one 6" diameter pipe is (neglect friction) ______ pipes of ______ diameter each. 13.____

A. 3; 3" B. 4; 3" C. 2; 4" D. 3; 4"

14. A roof tank is used to furnish the domestic water supply to a ten story building. This tank has a capacity of 5,900 gallons. At 10:00 A.M. one morning, the tank is half full. 14.____
If water is being used at the rate of 50 gals/min, the pump which is used to fill the tank has a rated capacity of 90 gals/min, the time it would take to fill the tank under these conditions is MOST NEARLY ______ hour(s), ______ minutes.

A. 2; 8 B. 1; 14 C. 2; 32 D. 1; 2

15. The number of gallons of water contained in a cylindrical swimming pool 8 feet in diameter and filled to a depth of 3 feet 6 inches is MOST NEARLY (assume 7.5 gallons = 1 cubic foot) 15.____

A. 30 B. 225 C. 1,320 D. 3,000

16. The charge for metered water is 52 1/2 cents per hundred cubic feet, with a minimum charge of $21 per annum. Of the following, the SMALLEST water usage in hundred cubic feet that would result in a charge GREATER than the minimum is 16.____

A. 39 B. 40 C. 41 D. 42

17. The annual frontage rent on a one-story building 40 ft. in length is $735.00. For each additional story, $52.50 per annum is added to the frontage rent. For demolition, the charge for wetting down is 3/8 of the annual frontage charge. 17.____
The charge for wetting down a building six stories in height, with a 40 ft. frontage, is MOST NEARLY

A. $369 B. $371 C. $372 D. $374

18. If the drawing of a piping layout is made to a scale of 1/4" equals one foot, then a 7'9" length of piping would be represented by a scaled length on the drawing of APPROXIMATELY _______ inches. 18.____

A. 2 B. 7 3/4 C. 23 1/4 D. 31

19. A plumbing sketch is drawn to a scale of eighth-size. A line measuring 3" on the sketch would be equivalent to _______ feet. 19.____

A. 2 B. 6 C. 12 D. 24

20. If 500 feet of pipe weighs 800 lbs., the number of pounds that 120 feet will weigh is MOST NEARLY 20.____

A. 190 B. 210 C. 230 D. 240

21. If a trench is excavated 3'0" wide by 5'6" deep and 50 feet long, the total number of cubic yards of earth removed is MOST NEARLY 21.____

A. 30 B. 90 C. 150 D. 825

22. Assume that a plumber earns $86,500 per year. 22.____
If eighteen percent of his pay is deducted for taxes and social security, his net weekly pay will be APPROXIMATELY

A. $1,326 B. $1,365 C. $1,436 D. $1,457.50

23. Assume that a plumbing installation is made up of the following fixtures and groups of fixtures: 12 bathroom groups each containing one W.C., one lavatory, and one bathtub with shower; 12 bathroom groups each containing one W.C., one lavatory, one bathtub, and one shower stall; 24 combination kitchen fixtures; 4 floor drains; 6 slop sinks without flushing rim; and 2 shower stalls (or shower bath). 23.____
The total number of fixtures for the above plumbing installation is MOST NEARLY

A. 60 B. 95 C. 120 D. 210

24. A triangular opening in a wall forms a 30-60 degree right triangle. 24.____
If the longest side measures 12'0", then the shortest side will measure

A. 3'0" B. 4'0" C. 6'0" D. 8'0"

25. You are directed to cut 4 pieces of pipe, one each of the following length: 2'6 1/4", 3'9 3/8", 4'7 5/8", and 5'8 7/8". 25.___
The total length of these 4 pieces is

A. 15'7 1/4" B. 15'9 3/8" C. 16'5 7/8" D. 16'8 1/8"

KEY (CORRECT ANSWERS)

1.	A	11.	D
2.	B	12.	D
3.	B	13.	B
4.	B	14.	B
5.	C	15.	C
6.	C	16.	C
7.	B	17.	D
8.	B	18.	A
9.	B	19.	A
10.	B	20.	A

21. A
22. B
23. C
24. C
25. D

SOLUTIONS TO PROBLEMS

1. 8'3 1/2" + x + x = 8'9 3/4" Then, 2x = 6 1/4", so x = 3 1/8"

2. 9/64 = .140625 = .1406

3. 1'2 3/16" + 1'5 1/2" +1'4 5/8" = 3'11 21/16" = 4'5/16"

4. 3 5/16" ÷ 1/8" =53/16 x 8/1 = 26.5. Then, (26.5)(1 ft.) = 26.5 feet

5. 50,631 - 39,842 = 10,789; 10,789 ÷ 100 = 107.89
Since the cost is .23 per 100 cubic feet or any fraction thereof, the cost will be (.23)(107) + .23 = $24.84

6. 9740 ÷ 1.04 = 9365 cu.ft.

7. 40 ÷ 7.5 = 5 1/3 cu.ft. of water per minute. The volume = (6)(5)(4) = 120 cu.ft. Thus, the number of minutes needed to fill the tank is 120 ÷ 5 1/3 = 22.5

8. 3" pipe: 3 x 7'6" = 22 1/2' x 14.5 lbs. = 326.25
4" pipe: 4 x 5' = 20' x 13 lbs. = 260
326.25 + 260 = 586.25 (most nearly 585)

9. Let x = pressure. Then, 34/12 = 14.7/x. So, 34x = 176.4
Solving, x $\approx$ 5 lbs./sq.in.

10. (8)(9)(63) = 4536 cu.ft. Since 1 cu.yd. = 27 cu.ft., 4536 cu.ft. is equivalent to 168 cu.yds.

11. Let x = correct amount. Then, $\frac{1}{1200} = \frac{1\frac{1}{3}}{x}$. Solving, x = 1600

12. 87,463 - 17,377 = 70,086; and 70,086 ÷ 100 = 700.86 $\approx$ 700 Then, (700)(.15) = $105.00

13. Cross-sectional area of a 6" diameter pipe = (π)(3")^2 = 9 π sq. in. Note that the combined cross-sectional areas of four 3" diameter pipes = (4) (π) (1.5")^2 = 9 π sq . in .

14. 90 - 50 = 40 gals/min. Then, 2950 ÷ 40 = 73.75 min. $\approx$ 1 hr. 14 min.

15. Volume = (π)(4)^2(3 1/2) = 56 π cu.ft. Then, (56 π)(7.5) = 1320 gals.

16. For 4100 cu.ft., the charge of (.525)(41) = $21,525 > $21

17. Rent = $73,500 + (5)($52.50) = $997,50. For demolition, the charge = (3/8)($997.50) $374

18. (1/4")(7.75) = 2"

19. (3")(8) = 24" = 2 ft.

20. Let x = weight. Then, 500/800 = 120/x . Solving, x = 192 190 lbs.

21. (3')(5 1/2')(50') = 825 cu.ft. Then, 825 ÷ 27 ≈ 30 cu.yds.

22. Net pay = (.82)($86,500) = $70,930/yr. Weekly pay = $70,930 ÷ 52 ≈ $1365

23. (12x3) + (12x4) +24+4+6+2= 120

24. The shortest side = (1/2)(hypotenuse) = (1/2)(12') = 6'

25. 2'6 1/4" + 3'9 3/8" + 4'7 5/8" + 5'8 7/8 " = 14'30 17/8" = 16'8 1/8"

TEST 2

DIRECTIONS: Each question or incomplete statement is followed by several suggested answers or completions. Select the one that BEST answers the question or completes the statement. *PRINT THE LETTER OF THE CORRECT ANSWER IN THE SPACE AT THE RIGHT.*

1. The sum of the following pipe lengths, 15 5/8", 8 3/4", 30 5/16" and 20 1/2", is 1.____

   A. 77 1/8" B. 76 3/16" C. 75 3/16" D. 74 5/16"

2. If the outside diameter of a pipe is 6 inches and the wall thickness is 1/2 inch, the inside area of this pipe, in square inches, is MOST NEARLY 2.____

   A. 15.7 B. 17.3 C. 19.6 D. 23.8

3. Three lengths of pipe 1'10", 3'2 1/2", and 5'7 1/2", respectively, are to be cut from a pipe 14'0" long. 3.____
   Allowing 1/8" for each pipe cut, the length of pipe remaining is

   A. 3'1 1/8" B. 3'2 1/2" C. 3'3 1/4" D. 3'3 5/8"

4. According to the building code, the MAXIMUM permitted surface temperature of combustible construction materials located near heating equipment is 76.5°C. (°F=(°Cx9/5)+32) 4.____
   Maximum temperature Fahrenheit is MOST NEARLY

   A. 170° F B. 195° F C. 210° F D. 220° F

5. A pump discharges 7.5 gals/minutes. 5.____
   In 2.5 hours the pump will discharge ______ gallons.

   A. 1125 B. 1875 C. 1950 D. 2200

6. A pipe with an outside diameter of 4" has a circumference of MOST NEARLY ______ inches. 6.____

   A. 8.05 B. 9.81 C. 12.57 D. 14.92

7. A piping sketch is drawn to a scale of 1/8" = 1 foot. 7.____
   A vertical steam line measuring 3 1/2" on the sketch would have an ACTUAL length of ______ feet.

   A. 16 B. 22 C. 24 D. 28

8. A pipe having an inside diameter of 3.48 inches and a wall thickness of .18 inches will have an outside diameter of ______ inches. 8.____

   A. 3.84 B. 3.64 C. 3.57 D. 3.51

9. A rectangular steel bar having a volume of 30 cubic inches, a width of 2 inches, and a height of 3 inches will have a length of ______ inches. 9.____

   A. 12 B. 10 C. 8 D. 5

10. A pipe weighs 20.4 pounds per foot of length. 10.____
    The total weight of eight pieces of this pipe with each piece 20 feet in length is MOST NEARLY ______ pounds.

    A. 460 B. 1,680 C. 2,420 D. 3,260

11. Assume that four pieces of pipe measuring 2'1 1/4", 4'2 3/4", 5'1 9/16", and 6'3 5/8", respectively, are cut with a saw from a pipe 20"0" long. Allowing 1/16" waste for each cut, the length of the remaining pipe is 11.___

A. 2'1 9/16" B. 2'2 9/16" C. 2'4 13/16" D. 2'8 9/16"

12. If one cubic inch of steel weighs 0.28 pounds, the weight, in pounds, of a steel bar 1/2" x 6" x 2'0" long is MOST NEARLY 12.___

A. 11 B. 16 C. 20 D. 24

13. If the circumference of a circle is equal to 31.416 inches, then its diameter, in inches, is equal to MOST NEARLY 13.___

A. 8 B. 9 C. 10 D. 13

14. Assume that a steam fitter's helper receives a salary of $171.36 a day for 250 days is considered a full work year. If taxes, social security, hospitalization, and pension deducted from his salary amounts to 16 percent of his gross pay, then his net yearly salary will be MOST NEARLY 14.___

A. $31,788 B. $35,982 C. $41,982 D. $42,840

15. If the outside diameter of a pipe is 14 inches and the wall thickness is 1/2 inch, then the inside area of the pipe, in square inches, is MOST NEARLY 15.___

A. 125 B. 133 C. 143 D. 154

16. A steam leak in a pipe line allows steam to escape at a rate of 50,000 pounds each month. Assuming that the cost of steam is $2.50 per 1,000 pounds, the TOTAL cost of wasted steam from this leak for a 12-month period would amount to 16.___

A. $125 B. $300 C. $1,500 D. $3,000

17. If 250 feet of 4" pipe weighs 400 pounds, the weight of this pipe per linear foot is _______ pounds. 17.___

A. 1.25 B. 1.50 C. 1.60 D. 1.75

18. A set of heating plan drawings is drawn to a scale of 1/4" = 1 foot. If a length of pipe measures 4 5/8" on the drawing, the ACTUAL length of the pipe, in feet, is 18.___

A. 16.3 B. 16.8 C. 17.5 D. 18.5

19. The TOTAL length of four pieces of pipe whose lengths are 3'4 1/2", 2'1 5/16", 4'9 3/8", and 2'3 1/4", respectively, is 19.___

A. 11'5 7/16" B. 11'6 7/16"
C. 12'5 7/16" D. 12'6 7/16"

20. Assume that a pipe trench is 3 feet wide, 3 feet deep, and 300 feet long. If the unit cost of excavating the trench is $120 per cubic yard, the TOTAL cost of excavating the trench is 20.___

A. $1,200 B. $12,000 C. $27,000 D. $36,000

21. The TOTAL length of four pieces of 1 1/2" galvanized steel pipe whose lengths are 7 ft. + 3 1/2 inches, 4 ft. + 2 1/4 inches, 6 ft. + 7 inches, and 8 ft. +5 1/8 inches is 21.____

A. 26 feet + 5 7/8 inches B. 25 ft. + 6 7/8 inches
C. 25 feet + 4 1/4 inches D. 25 ft. + 3 3/8 inches

22. A swimming pool is 25' wide by 75' long and has an average depth of 5'. 1 cubic foot contains 7.5 gallons of water. The capacity, when filled to the overflow, is ______ gallons. 22.____

A. 9,375 B. 65,625 C. 69,005 D. 70,312

23. The sum of 3 1/4, 5 1/8, 2 1/2 , and 3 3/8 is 23.____

A. 14 B. 14 1/8 C. 14 1/4 D. 14 3/8

24. Assume that it takes 6 men 8 days to do a particular job. If you have only 4 men available to do this job and they all work at the same speed, then the number of days it would take to complete the job would be 24.____

A. 11 B. 12 C. 13 D. 14

25. The total length of four pieces of 2" O.D. pipe, whose lengths are 7'3 1/2", 4'2 3/16", 5'7 5/16", and 8'5 7/8", respectively, is MOST NEARLY 25.____

A. 24'6 3/4" B. 24'7 15/16"
C. 25'5 13/16" D. 25'6 7/8"

KEY (CORRECT ANSWERS)

1.	C	11.	B
2.	C	12.	C
3.	D	13.	C
4.	A	14.	B
5.	A	15.	B
6.	C	16.	C
7.	D	17.	C
8.	A	18.	D
9.	D	19.	D
10.	D	20.	B

21. A
22. D
23. C
24. B
25. D

SOLUTIONS TO PROBLEMS

1. 15 5/8" + 8 3/4" + 30 5/16" + 20 1/2" = 73 35/16" = 75 3/16"
2. Inside diameter = 6" - 1/2" - 1/2" = 5". Area = $(\pi)(5/2")^2 \approx 19.6$ sq. in.
3. Pipe remaining = 14' - 1'10" - 3'2 1/2" - 5'7 1/2" - (3)(1/8") = 3'3 5/8"
4. 76.5 x 9/5 = 137.7 + 32 = 169.7
5. 7.5 x 150 = 1125
6. Radius = 2" Circumference = $(2\pi)(2") \approx 12.57"$
7. 3 1/2" 1/8" = (7/2)(8/1) = 28 Then, (28)(1 ft.) = 28 feet
8. Outside diameter = 3.48" + .18" + .18" = 3.84"
9. 30 = (2)(3)(length). So, length = 5"
10. Total weight = (20.4)(8)(20) $\approx$ 3260 lbs.
11. 20' - 2'1 1/4" - 4'2 3/4" - 5'1 9/16" - 6'3 5/8" - (4)(1/16") = 2'2 9/16"
12. Weight = (.28)(1/2")(6")(24") = 20.16 $\approx$ 20 lbs.
13. Diameter = 31.416" $\div \pi \approx$ 10"
14. His net pay for 250 days = (.84)($171.36)(250) = $35,985.60 ≈ $35,928 (from answer key)
15. Inside diameter = 14" - 1/2" - 1/2" = 13". Area = $(\pi)(13/2")^2 \approx$ 133 sq.in
16. (50,000 lbs.)(12) = 600,000 lbs. per year. The cost would be ($2.50)(600) = $1500
17. 400 $\div$ 250 = 1.60 pounds per linear foot
18. 4 5/8" $\div$ 1/4" = 37/8 . 4/1 = 18.5 Then, (18.5)(1 ft.) = 18.5 feet
19. 3'4 1/2" + 2'1 5/16" + 4'9 3/8" + 2'3 1/4" = 11'17 23/16" = 12'6 7/16"
20. (3')(3')(300') = 2700 cu.ft., which is 2700 $\div$ 27 = 100 cu.yds. Total cost = ($120)(100) = $12,000
21. 7'3 1/2" + 4'2 1/4" + 6'7" + 8'5 1/8" = 25'17 7/8" = 26'5 7/8"
22. (25)(75)(5) = 9375 cu.ft. Then, (9375)(7.5) ≈ 70,312 gals.
23. 3 1/4 + 5 1/8 + 2 1/2 + 3 3/8 = 13 10/8 = 14 1/4
24. (6) (8) = 48 man-days. Then, 48 $\div$ 4 = 12 days
25. 7'3 1/2" + 4'2 3/16" + 5'7 5/16" + 8'5 7/8"= 24'17 30/16" = 25'6 7/8"

TEST 3

DIRECTIONS: Each question or incomplete statement is followed by several suggested answers or completions. Select the one that BEST answers the question or completes the statement. *PRINT THE LETTER OF THE CORRECT ANSWER IN THE SPACE AT THE RIGHT.*

1. The time required to pump 2,500 gallons of water out of a sump at the rate of 12 1/2 gallons per minutes would be ______ hour(s) ______ minutes. 1.____

   A. 1; 40 B. 2; 30 C. 3; 20 D. 6; 40

2. Copper tubing which has an inside diameter of 1 1/16" and a wall thickness of .095" has an outside diameter which is MOST NEARLY ______ inches. 2.____

   A. 1 5/32 B. 1 3/16 C. 1 7/32 D. 1 1/4

3. Assume that 90 gallons per minute flow through a certain 3-inch pipe which is tapped into a street main. 3.____
   The amount of water which would flow through a 1-inch pipe tapped into the same street main is MOST NEARLY ______ gpm.

   A. 90 B. 45 C. 30 D. 10

4. The weight of a 6 foot length of 8-inch pipe which weighs 24.70 pounds per foot is ______ lbs. 4.____

   A. 148.2 B. 176.8 C. 197.6 D. 212.4

5. If a 4-inch pipe is directly coupled to a 2-inch pipe and 16 gallons per minute are flowing through the 4-inch pipe, then the flow through the 2-inch pipe will be ______ gallons per minute. 5.____

   A. 4 B. 8 C. 16 D. 32

6. If the water pressure at the bottom of a column of water 34 feet high is 14.7 pounds per square inch, the water pressure at the bottom of a column of water 18 feet high is MOST NEARLY ______ pounds per square inch. 6.____

   A. 8.0 B. 7.8 C. 7.6 D. 7.4

7. If there are 7 1/2 gallons in a cubic foot of water and if water flows from a hose at a constant rate of 4 gallons per minute, the time it should take to COMPLETELY fill a tank of 1,600 cubic feet capacity with water from that hose is ______ hours. 7.____

   A. 300 B. 150 C. 100 D. 50

8. Each of a group of fifteen water meter readers read an average of 62 water meters a day in a certain 5-day work week. A total of 5,115 meters are read by this group the following week. 8.____
   The TOTAL number of meters read in the second week as compared to the first week shows a

   A. 10% increase B. 15% increase
   C. 20% increase D. 5% decrease

9. A certain water consumer used 5% more water in 1994 than he did in 1993. If his water consumption for 1994 was 8,375 cubic feet, the amount of water he consumed in 1993 was MOST NEARLY ______ cubic feet. 9.___

A. 9,014 B. 8,816 C. 7,976 D. 6,776

10. Assume that a water meter reads 40,175 cubic feet and that the previous reading was 29,186 cubic feet. If the charge for water is 92 cents per 100 cubic feet or any fraction thereof, the bill for the amount of water used since the previous meter reading should be 10.___

A. $100.28 B. $101.04 C. $101.08 D. $101.20

11. A leaking faucet caused a loss of 216 cubic feet of water in a 30-day month. If there are 7.5 gallons in a cubic foot of water, then the AVERAGE loss of water per hour for that month was ______ gallons. 11.___

A. 2 1/4 B. 2 1/8 C. 2 D. 1 3/4

12. The fraction which is equal to .375 is 12.___

A. 3/16 B. 5/32 C. 3/8 D. 5/12

13. A square backyard swimming pool, each side of which is 10 feet long, is filled to a depth of 3 1/2 feet. If there are 7 1/2 gallons in a cubic foot of water, the number of gallons of water in the pool is MOST NEARLY ______ gallons. 13.___

A. 46.7 B. 100 C. 2,625 D. 3,500

14. When 1 5/8, 3 3/4, 6 1/3, and 9 1/2 are added, the resulting sum is 14.___

A. 21 1/8 B. 21 1/6 C. 21 5/24 D. 21 1/4

15. When 946 1/2 is subtracted from 1,035 1/4, the result is 15.___

A. 87 1/4 B. 87 3/4 C. 88 1/4 D. 88 3/4

16. When 39 is multiplied by 697, the result is 16.___

A. 8,364 B. 26,283 C. 27,183 D. 28,003

17. When 16.074 is divided by .045, the result is 17.___

A. 3.6 B. 35.7 C. 357.2 D. 3,572

18. To dig a trench 3'0" wide, 50'0" long, and 5'6" deep, the total number of cubic yards of earth to be removed is MOST NEARLY 18.___

A. 30 B. 90 C. 140 D. 825

19. The TOTAL length of four pieces of 2" pipe, whose lengths are 7'3 1/2", 4'2 3/16", 5'7 5/16", and 8'5 7/8", respectively, is 19.___

A. 24'6 3/4" B. 24'7 15/16"
C. 25'5 13/16" D. 25'6 7/8"

20. A hot water line made of copper has a straight horizontal run of 150 feet and, when installed, is at a temperature of 45° F. In use, its temperature rises to 190° F. If the coefficient of expansion for copper is 0.0000095" per foot per degree F, the TOTAL expansion, in inches, in the run of pipe is given by the product of 150 multiplied by 0.0000095 by 20.____

A. 145
B. 145 x 12
C. 145 divided by 12
D. 145 x 12 x 12

21. A water storage tank measures 5' long, 4' wide, and 6' deep and is filled to the 5 1/2' mark with water. If one cubic foot of water weighs 62 pounds, the number of pounds of water required to COMPLETELY fill the tank is 21.____

A. 7,440 B. 6,200 C. 1,240 D. 620

22. Assume that a pipe worker earns $83,125.00 per year. If seventeen percent of his pay is deducted for taxes, social security, and pension, his net weekly pay will be APPROXIMATELY 22.____

A. $1598.50 B. $1504.00 C. $1453.00 D. $1325.00

23. If eighteen feet of 4" cast iron pipe weighs approximately 390 pounds, the weight of this pipe per lineal foot will be MOST NEARLY ______ lbs. 23.____

A. 19 B. 22 C. 23 D. 25

24. If it takes 3 men 11 days to dig a trench, the number of days it will take 5 men to dig the same trench, assuming all work is done at the same rate of speed, is MOST NEARLY 24.____

A. 6 1/2 B. 7 3/4 C. 8 1/4 D. 8 3/4

25. If a trench is dug 6'0" deep, 2'6" wide, and 8'0" long, the area of the opening, in square feet, is MOST NEARLY 25.____

A. 48 B. 32 C. 20 D. 15

KEY (CORRECT ANSWERS)

1. C
2. D
3. D
4. A
5. B

6. B
7. D
8. A
9. C
10. D

11. A
12. C
13. C
14. C
15. D

16. C
17. C
18. A
19. D
20. A

21. D
22. D
23. B
24. A
25. C

SOLUTIONS TO PROBLEMS

1. 2500 ÷ 12 1/2 = 200 min. = 3 hrs. 20 min.

2. 1 1/16" + .095" + .095" = 1.0625 + .095 + .095 = 1.2525" ≈ 1 1/4"

3. Cross-sectional areas for a 3-inch pipe and a 1-inch pipe are $(\pi)(1.5)^2$ and $(\pi)(.5)^2$ = 2.25 π and .25 π, respectively. Let x = amount of water flowing through the 1-inch pipe. Then, $\frac{90}{x} = \frac{2.25\pi}{.25\pi}$. Solving, x = 10 gals/min

4. (24.70)(6) = 148.2 lbs.

5. $\frac{\text{4" pipe}}{\text{16 gallons}} = \frac{\text{2" pipe}}{\text{x gallons}}$, 4x = 32, x = 8

6. Let x = pressure. Then, 34/18 = 14.7/x . Solving, x ≈ 7.8

7. (1600)(7.5) = 12,000 gallons. Then, 12,000 ÷ 4 = 3000 min. = 50 hours

8. (15)(62)(5) = 4650. Then, (5115-4650)/4650 = 10% increase

9. 8375 ÷ 1.05 ≈ 7976 cu.ft.

10. 40,175 - 29,186 = 10,989 cu.ft. Then, 10,989 100 = 109.89. Since .92 is charged for each 100 cu.ft. or fraction thereof, total cost = (.92)(110) = $101.20

11. (216)(7.5) = 1620 gallons. In 30 days, there are 720 hours. Thus, the average water loss per hour = 1620 ÷ 720 = 2 1/4 gallons.

12. .375 = 375/1000 = 3/8

13. Volume = (10)(10)(3 1/2) = 350 cu.ft. Then, (350)(7 1/2) = 2625 gallons

14. 1 5/8 + 3 3/4 + 6 1/3 + 9 1/2 = 19 53/24 = 21 5/24

15. 1035 1/4 - 946 1/2 = 88 3/4

16. (39)(697) = 27,183

17. 16.074 .045 = 357.2

18. (3')(50')(5 1/2') = 825 cu.ft. ≈ 30 cu.yds., since 1 cu.yd. = 27 cu.ft.

19. 7'3 1/2" + 4'2 3/16" + 5'7 5/16" + 8'5 7/8" = 24'17 30/16" = 25'6 7/8"

20. Total expansion = (150)(.0000095)(145)

21. Number of pounds needed = (5) (4)(6-5 1/2)(62) = 620

22. Net annual pay = (\$83,125)(.83) ≈ \$69000. Then, the net weekly pay = \$69000 ÷ 52 ≈ \$1325 (actually about \$1327)

23. 390 lbs. ÷ 18 = 21.6 lbs. per linear foot

24. (3)(11) = 33 man-days. Then, 33 ÷ 5 = 6.6 ≈ 6 1/2 days

25. Area = (8')(2 1/2') = 20 sq.ft.

ARITHMETIC OF SEWAGE TREATMENT

The English system of measurements is used for computations at sewage treatment works, except in the case of a few determinations. The metric system will be mentioned where the metric units are used.

Basic Units

Linear	1 inch (in.)	= 2.540 centimeters (cm)
	1 foot (ft.)	= 12 inches (in.)
	1 yard (yd.)	= 3 feet (ft.)
	1 mile	= 5,280 feet
	1 meter (m)	= 39.37 in. = 3.281 ft. = 1.094 yd.
	1 meter	= 100 centimeters
Area	1 square foot (sq. ft.)	= 144 square inches (sq. in.)
	1 square yard (sq. yd.)	= 9 sq. ft.
	1 acre	= 43,560 sq. ft.
	1 square mile	= 640 acres
Volume	1 cubic foot	= 1728 cubic inches (cu. in.)
	1 cubic yard	= 27 cu. ft.
	1 cubic foot	= 7.48 gallons
	1 gallon (gal.)	= 231 cu. in.
	1 gallon	= 4 quarts (qt)
	1 gallon	= 3.785 liters(1)
	1 liter	= 1000 milliliters (ml)
Weight	1 pound (lb.)	= 16 ounces = 7000 grains = 453.6 grams
	1 ounce	= 28.35 grams (g)
	1 kilogram	= 1000 grams
	1 gram	=1000 milligrams (mg)
	1 cu. ft. water	= 62.4 pounds
	1 gallon water	= 8.33 pounds
	1 liter water	= 1 kilogram
	1 milliliter water	= 1 gram

Definition of Terms

A *ratio* is the indicated division of two pure numbers. As such is indicates the relative magnitude of two quantities. The ratio of 2 to 3 is written 2/3.

A *pure* number is used without reference to any particular thing.

A *concrete* number applies to a particular thing and is the product of a pure number and a physical unit. 5 ft. means 5 times 1 ft. or 5 X (1 ft.).

Rate units are formed when one physical unit is divided by another.

$$\frac{60\text{ft.}}{2\text{sec.}} = 30\frac{(\text{ft.})}{(\text{sec.})}$$

Physical units can be formed by multiplying two or more other physical units.

1 ft. X 1 ft. = 1 ft. X ft. = 1 ft.^{2} (square foot)

Physical units may cancel each other.

$$\frac{6 \text{ ft X } 7.48 \text{ gallons}}{1 \text{ft.}} = 6 \text{ X } 7.48 \text{ gallons}$$

Per cent means per 100 and is the numerator of a fraction whose denominator is always 100. It may be expressed by the symbol "%". The word *per* refers to a fraction whose numerator precedes *per* and whose denominator follows. Hence "per" means "divided by." It is often indicated by a sloping line as "/."

Problem: What is 15 per cent of 60?

$$60 \text{X} \frac{15}{100} = \frac{900}{100} = 9$$

Problem: One pound of lime is stirred into one gallon of water.

What is the per cent of lime in the slurry ?

$$\frac{1}{1+8.33} \text{X} 100 = \frac{100}{1+8.33} = 10.7 \text{ per cent}$$

Formulas

Circumference of a circle = $\Pi D = 2\Pi R$

Area of a circle $= \Pi R^2 = \frac{\Pi D^2}{4}$

$\Pi = 3.1416$
Area of triangle = 1/2 base X altitude
Area of rectangle = base X altitude
Cylindrical area = circumference of base X length
Volume of cylinder = area of base X length
Volume of rectangular tank = area of bottom X depth
Volume of cone = 1/3 X area of base X height
Velocity = distance divided by time. Inches, feet, or miles divided by hours, minutes, or seconds.
Discharge = volume of flow divided by time.
Gallons or cubic feet divided by days, hours, minutes, or seconds.
1 cu. ft. per sec. = 647,000 gallons per day.
1 mgd = 1.54 cfs = 92.4 cfm

Detention Time. The theoretical time equals the volume of tank divided by the flow per unit time. The flow volume and tank volume must be in the same units.

$$\frac{20{,}000 \text{ gal}}{200 \frac{\text{gal}}{\text{min.}}} = 100 \text{ minutes}$$

Problem: A tank is 60 X 20 X 30 ft. The flow is 5 mgd.

What is the detention time in hours?

1 mgd = 92.4 cfm

$$\frac{60\text{ ft.} \times 20\text{ ft.} \times 30\text{ ft.}}{92.4 \times 5\ \frac{\text{ft}^3}{\text{min}}} = 78 \text{ min. or 1 hr. and 18 min. or 1.3 hours}$$

Surface Settling Rate:

This means gallons per square foot of lank surface per day.

Problem: If the daily flow is 0.5 mgd and the tank is 50 ft. long and 12 ft. wide, calculate the surface settling rate.

$$\frac{500{,}000\text{ gal./day}}{50\text{ ft.} \times 12\text{ ft.}} = \frac{833\text{ gal.}}{\text{ft.}^2 \times \text{day}}$$

Weir Overflow Rate:

This means gallons per day per foot length of weir.

Problem: A circular settling tank is 90 ft. in diameter. The flow is 3.0 mgd. Calculate the weir overflow rate.

$$\frac{3{,}000{,}000\text{ gal./day}}{\Pi \times 90\text{ ft.}} = \frac{10{,}600\text{ gal}}{\text{ft.} \times \text{day}}$$

Rate of Filtration: The mgd is divided by the acres of stone to give

$$\frac{\text{mg}}{\text{acre} \times \text{day}} = \text{mgad}$$

$$\frac{\text{mg}}{\text{acre} \times \text{ft.} \times \text{day}} = \text{mgaftd}$$

An acre-ft. is an acre in area and 1 ft. deep.
A fixed-nozzle filter is 140 x 125 feet. Stone is six feet deep. Flow is 9 mgd. Calculate the rate of dosing or hydraulic loading in mg per acre-foot per day.

$$\frac{140 \times 125}{43560} = \text{acres} = 0.402$$

$$0.402 \times 6 = 2.412 \text{ acre-feet}$$

$$\frac{9}{2.412} = \frac{\text{mg}}{\text{acre} \times \text{ft.} \times \text{day}} = 3.73$$

The BOD of a settling tank effluent is 200 ppm. If 15 lb. of BOD per 1000 ft.^{3} of stone is to be the organic loading, how many cubic feet of stone are necessary with a hydraulic loading of 3 mgd.

$$\frac{200 \times 8.33 \times 3 \times 1000}{15} = 333{,}333 \text{ ft.}^3$$

$$\frac{333,333}{6} = 55,500 \text{ ft.}^2 \text{ for filter area if stone is 6 ft. deep.}$$

Parts per million:

This is a weight ratio. Any unit may be used; pounds per million pounds or milligrams per liter if the liquid has a specific gravity equal to water or very nearly so. 1 liter of water = 1,000,000 milligrams.

1 ppm = 8.33 lbs. per million gallons
1 ppm = 1 milligram per liter

A sewage with 600 ppm suspended solids has 600 X 8.33 = 4998 lb. of suspended solids per million gallons.

Efficiency of Removal:

$$\frac{\text{ppm influent - ppm effflueny}}{\text{ppm influent}} 100 = \text{percent efficiency of removal}$$

Percent of Moisture:

$$\frac{\text{wt. of wet sludge - wt. of day sludge}}{\text{wt. of wet sludge}} 100 = \text{percent moisture}$$

Percent of Dry solids:

$$\frac{\text{wt. of day sludge}}{\text{wt. of wet sludge}} 100 = \text{parcent day solids}$$

Other calculated quantities that need no special explanation are:
Square feet of sludge drying bed per capita
Cubic feet of digestion space per capita
Cubic feet of sludge produced per day per capita
Cubic feet of grit per million gallons
Pounds of sludge per capita per day
Cubic feet of gas per capita per day
Kilowatt-hours per million gallons pumped

Specific Gravity: This is the ratio of the density of a substance to the density of water. There is no unit. Density = the weight of unit volume.

$$\text{S.G.} = \frac{\text{(wt. bottle with sludge) - (wt. of empty bottle)}}{\text{(wt. bottle with water) - (wt. of empty bottle)}}$$

1 gallon of water = 8.33 lbs.
1 cu. ft. of water = 62.4 lbs.
These vary slightly with temperature.
Water at 32° F. = 62.417 lb./ft.^3

Water at 62° F. = 62.355 lb./ft.^{3}

Water at 212° F. = 59.7 lb./ft.^{3}

Ice = 57.5 lb./ft.^{3}

Problem: What is the weight of dry solids in 1000 gallons of 10% sludge whose specific gravity is 1.04?

$$1000 \times 8.33 \times 1.40 \times \frac{10}{100} = 866.3 \text{ lbs.}$$

Mixtures:

If two materials of different percentages are to be mixed to produce an intermediate percentage, it may be done by rectangle method. Problem: We have 30 per cent and 50 per cent material. In what ratio shall they be mixed to produce 37 per cent material.

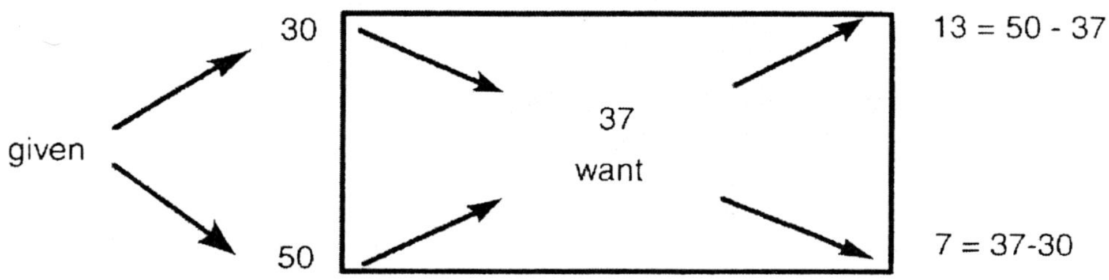

Desired ration is 13 parts of the 30 per cent and 7 parts of the 50 per cent. This will give us 20 parts of 37 per cent.

GLOSSARY OF SEWAGE TREATMENT TERMS

CONTENTS

	Page
Activated Sludge Process ... Arrester, Flame	1
Bacteria ... Burner, Waste Gas	2
Centrifuge ... Chlorine	3
Clarifier ... Digestion	4
Dilution ... Effluent	5
Ejector, Pneumatic ... Floc	6
Flocculator ... Grease	7
Grinder, Screenings ... Impeller	8
Index, Sludge Volume ... Mold	9
Most Probable Number, (MPN) ... Period	10
pH ... Pump	11
Purification ... Rate	12
Reaeration ... Sedimentation	13
Seeding, Sludge ... Sewer	14
Sewerage ... Solids	15
Squeegee ... Treatment	16
Trap, Flame ... Zooglea	17

GLOSSARY OF SEWAGE TREATMENT TERMS

A

Activated Sludge Process.—See **Process, Activated Sludge.**

Acre-Foot.—A unit of volume used to express the amount of material in a trickling filter. A depth of one foot on an area of one acre is an acre-foot. Regardless of shape, 43,560 cubic feet is equivalent to one acre foot.

Adsorption.—The adherence of dissolved, colloidal, or finely divided solids on the surfaces of solid bodies with which they are brought into contact.

Aeration.—The bringing about of intimate contact between air and a liquid by one of the following methods: Spraying the liquid in the air; or by agitation of the liquid to promote surface absorption of air.

Diffused Air.—Aeration produced in a liquid by air passed through a diffuser.

Mechanical.— (1) The mixing, by mechanical means, of sewage and activated sludge, in the aeration tank of the activated sludge process, to bring fresh surfaces of liquid into contact with the atmosphere. (2) The introduction of atmospheric oxygen into a liquid by the mechanical action of paddle or spray mechanisms.

Modified.—A modification of the activated sludge process in which a shortened period of aeration is employed with a reduced quantity of suspended solids in the mixed liquor.

Paddle-Wheel.—The mechanical agitation of sewage in the aeration tanks of the activated sludge process by means of paddle wheels.

Spiral Flow.—A method of diffusing air in an aeration tank of the activated sludge process, where, by means of properly designed baffles, and the proper location of diffusers, a spiral or helical movement is given to the air and the tank liquor.

Stage.—Division of activated sludge treatment into stages with intermediate settling tanks and return of sludge in each stage.

Step.—A procedure for adding increments of sewage along the line of flow in the aeration tanks of an activated sludge plant.

Tapered.—The method of supplying varying amounts of air into the different parts of an aeration tank in the activated sludge process, more at the inlet, less near the outlet, and approximately proportional to the oxygen demand of the mixed liquor under aeration.

Algae.—Primitive plants, one or many-celled, usually aquatic and capable of elaborating their foodstuffs by photosynthesis.

Algicide.—Any substance which kills algae.

Alkaline.—Water or soils containing sufficient amounts of alkaline substances to raise the pH above 7.0, or to harm the growth of crops.

Alkalinity.—A term used to represent the content of carbonates, bicarbonates, hydroxides, and occasionally borates, silicates, and phosphates in water. It is expressed in parts per million of calcium carbonate.

Alum.—A common name for aluminum sulfate.

Arrester, Flame.—A safety device on a gas line which allows gas, but not a flame, to pass through.

B

Bacteria.—Primitive plants, generally free of pigment, which reproduce by dividing in one, two, or three planes. They occur as single cells, groups, chains, or filaments, and do not require light for their life processes. They may be grown by special culturing out of their native habitat.

Aerobic.—Bacteria which require free (elementary) oxygen for their growth.

Anaerobic.—Bacteria which grow in the absence of free oxygen and derive oxygen from breaking down complex substances.

Coli-Aerogenes.—See Bacteria, Coliform Group.

Coliform Group.—A group of bacteria, predominantly inhabitants of the intestine of man but also found on vegetation, including all aerobic and facultative anaerobic grain-negative, non-spore-forming bacilli that ferment lactose with gas formation. This group includes five tribes of which the very great majority are Eschericheae. The Eschericheae tribe comprises three genera and ten species, of which *Escherichia Coli* and *Aerobacter Aerogenes* are dominant. *The Escherichia Coli* are normal inhabitants of the intestine of man and all vertebrates whereas Aerobacter Aerogenes normally are found on grain and plants, and only to a varying degree in the intestine of man and animals. Formerly referred to as *B.Coli*, B.Coli group, *ColiAerogenes Group.*

Facultative Anaerobic.—Bacteria which can adapt themselves to growth in the presence, as well as in the absence, of uncombined oxygen.

Parasitic.—Bacteria which thrive on other living organisms.

Pathogenic.—Bacteria which can cause disease.

Saprophytic.—Bacteria which thrive upon dead organic matter.

Bacterial Count.—A measure of the concentration of bacteria.

Most Probable Number.—See Page 10.

Plate.—Number of colonies of bacteria grown on selected solid media at a given temperature and incubation period, usually expressed as the number of bacteria per milliliter of sample.

Bed, Sludge.—An area comprising natural or artificial layers of porous material upon which digested sewage sludge is dried by drainage and evaporation. A sludge bed may be opened to the atmosphere or covered usually with a greenhouse-type superstructure. Also called Sludge Drying Bed.

Biochemical.—Resulting from biologic growth or activity, and measured by or expressed in terms of the ensuing chemical change.

Biochemical Action.—Chemical changes resulting from the metabolism of living organisms.

Biochemical Oxygen Demand (BOD).—The quantity of oxygen utilized in the biochemical oxidation of organic matter in a specified time and at a specified temperature. It is not related to the oxygen requirements in chemical combustion, being determined entirely by the availability of the material as a biological food and by the amount of oxygen utilized by the microorganisms during oxidation.

Biochemical Oxygen Demand, Standard.—Biochemical oxygen demand as determined under standard laboratory procedure for five days at 20°C, usually expressed in parts per million.

Buffer.—The action of certain solutions in opposing a change of composition, especially of hydrogen-ion concentration.

Burner, Waste Gas.—A device in a sewage treatment plant for burning the waste gas from a sludge-digestion tank.

C

Centrifuge.—A mechanical device utilizing centrifugal force to separate solids from liquids or for separating liquid emulsions.

Chamber.—A general term applied to a space enclosed by walls or to a compartment, often prefixed by a descriptive word, such as "grit chamber," "screen chamber," "discharge chamber," or "flushing chamber," indicating its function.

Chloramines.—Compounds of organic amines or inorganic ammonia with chlorine.

Chloride of Lime.—Obsolete term; see Chlorinated Lime.

Chlorinated Lime.—A combination of slaked lime and chlorine gas (also termed Bleaching Powder, Chloride of Lime, Hypochlorite of Lime, etc.). When dissolved in water, it serves as a source of chlorine.

Chlorination.—The application of chlorine.

Break-Point.—The application of chlorine to water, sewage or industrial wastes containing free ammonia to provide free residual chlorination.

Post.—The application of chlorine to water, sewage, or industrial wastes subsequent to any treatment. The term refers only to a point of application.

Pre.—The application of chlorine to water, sewage, or industrial wastes prior to any treatment. This term refers only to a point of application.

Chlorine.—An element, when uncombined, exists as a greenish yellow gas about 2.5 times as heavy as air. Under atmospheric pressure and at a temperature of —30.1°F the gas becomes an amber liquid about 1.5 times as heavy as water. The chemical symbol of chlorine is Cl, UP atomic weight is 35.457, and its molecular weight is 70.914.

Available.—A term used in rating chlorinated lime and hypochlorites as to their total oxidizing power.

Combined Available Residual.—That portion of the total residual chlorine remaining in water, sewage, or industrial wastes at the end of a specified contact period, which will react chemically and biologically as chloramines, or organic chloramines.

Demand.—The difference between the amount of chlorine added to water, sewage, or industrial wastes and the amount of residual chlorine remaining at the end of a specified contact period. The demand for any given water varies with the amount of chlorine applied, time of contact, and temperas pure.

Dose.—The amount of chlorine applied to a liquid, usually expressed in parts per million, or pounds per million gallons.

Free Available Residual.—That portion of the total residual chlorine remaining in water. sewage, or industrial wastes at the end of a specified contact period. which will react chemically and biologically as hypochlorous acid, hypochlorite ion, or molecular chlorine.

Liquid.—An article of commerce. Chlorine gas is generally manufactured by the electrolysis of a solution of common salt. The gas is dried and purified and is then liquefied by a combination of compression and refrigeration. Liquid chlorine is shipped under pressure in steel containers.

Residual.--The total amount of chlorine (combined and free available chlorine) remaining in water, sewage, or industrial wastes at the end of a specified contact period following chlorination.

Test, Iodometric.—The determination of residual chlorine in water, sewage, or industrial wastes by adding potassium iodide and titrating the liberated iodine with a standard solution of sodium thiosulfate, using starch solution as a colorimetric indicator.

Test, Ortho-Tolidine.—The determination of residual chlorine in water, sewage, or industrial wastes, using ortho-tolidine reagent and colorimetric standards.

Clarifier.—See Tank, Sedimentation.

Coagulation.—(1) The agglomeration of colloidal or finely divided suspended matter by the addition to the liquid of an appropriate chemical coagulant, by biological processes, or by other means. (2) The process of adding a coagulant and the necessary reacting chemicals.

Coils, Digester.—A system of pipes for hot water or steam installed in a sludge-digestion tank for the purpose of heating the sludge.

Coli-Aerogenes, or Coliform Group.—See Bacteria, Coliform Group.

Collector, Grit.—A device placed in a grit chamber to convey deposited grit to one end of the chamber for removal.

Scum.—A mechanical device for skimming and removing scum from the surface of settling tanks.

Sludge.—A mechanical device for scraping the sludge on the bottom of a settling tank to a sump, from which it can be drawn by hydrostatic or mechanical action.

Colloids.—Finely divided solids which will not settle but may be removed by coagulation or biochemical action.

Comminution.—The process of screening sewage and cutting the screenings into particles sufficiently fine to pass through the screen openings.

Concentration, Hydrogen-Ion.—See pH.

Copperas.—A common name for ferrous sulfate.

Copperas, Chlorinated.—A solution of ferrous sulfate and ferric chloride produced by chlorinating a solution of ferrous sulfate.

Cross Connection.—In plumbing, a physical connection through which a supply of potable water could be contaminated, polluted, or infected. A physical connection between water supplies from different systems.

Cubic Foot per Second.—A unit of discharge for measurement of flowing liquid, equal to a flow of one cubic foot per second past a given section. Also called Second-Foot.

D

Decomposition of Sewage.—The breakdown of the organic matter in sewage through aerobic and anaerobic processes.

Denitrification.—The reduction of nitrates in solution by biochemical action.

Deoxygenation.—The depletion of the dissolved oxygen in a liquid. Under natural conditions associated with the biochemical oxidation of organic matter present.

Detritus.—The sand, grit, and other coarse material removed by differential sedimentation in a relatively short period of detention.

Diffuser.—A porous plate or tube through which air is forced and divided into minute bubbles for diffusion in liquids. Commonly made of carborundum, alundum, or silica sand.

Digester.—A tank in which the solids resulting from the sedimentation of sewage are stored for the purpose of permitting anaerobic decomposition to the point of rendering the product nonputrescible and inoffensive. Erroneously termed digestor.

Digestion.—The processes occurring in a digester.

Mesophilic.—Digestion by biological action at or below 113°F.

Separate Sludge.—The digestion of sludge in separate tanks in which it is placed after it has been allowed to settle in other tanks.

Single-Stage Sludge.—Sludge digestion limited to a single tank for the entire digestion period.

Stage.—The digestion of sludge progressively in several tanks arranged in series.

Thermophilic.—Digestion carried on at a temperature generally between 113°F and 145°F.

Dilution. — (1) A method of disposing of sewage, industrial waste, or sewage treatment plant effluent by discharging it into a stream or body of water. (2) The ratio of volume of flow of a stream to the total volume of sewage or sewage treatment, ant effluent discharged into it.

Disinfection.—The killing of the larger portion (but not necessarily all) of the harmful and objectional microorganisms in, or on, a medium by means of chemicals, heat, ultraviolet light, etc.

Distributor.—A device used to apply liquid to the surface of a filter or contact bed, of two general types, fixed o movable. The fixed type may consist of perforated pipes or notched troughs, sloping boards, or sprinkler nozzles. The movable type may consist of rotating disks or rotating, reciprocating, or traveling perforated pipes or troughs applying a spray, or a thin sheet of liquid.

Dosing Tank.—A tank into which raw or partly treated sewage is introduced and held until the desired quantity has been accumulated, after which it is discharged at such a rate as may be necessary for the subsequent treatment.

Dryer.—A device utilizing heat to remove water.

Flash.—A device for vaporizing water from partly dewatered and finely divided sludge through contact with a current of hot gas or superheated vapor. Included is a squirrel cage mill for separating the sludge cake into fine particles.

Rotary.—A long steel cylinder, slowly revolving, with its long axis slightly inclined, through which passes the material to be dried in hot air. The material passes through from inlet to outlet, tumbling about.

E

E. Coli.—(Escherichia Coli).—A species of genus Escherichia bacteria, normal inhabitant of the intestine of man and all vertebrates. This species is classified among the Coliform Group. See Bacteria, Coliform Group.

Efficiency.—The ratio of the actual performance of a device to the theoretically perfect performance usually expressed as a percentage.

Average.—The efficiency of a machine or mechanical device over the range of load through which the machine operates.

Filter.—The operating results from a filter as measured by various criteria such as percentage reduction in suspended matter, total solids, biochemical oxygen demand, bacteria, color, etc.

Pump.—The ratio of energy converted into useful work to the energy applied to the pump shaft, or the energy difference in the water at the discharge and suction nozzles divided by the energy input at the pump shaft.

Wire-to-Water.—The ratio of the mechanical output of a pump, to the electrical input at the meter.

Effluent.—(1) A liquid which flows out of a containing space. (2) Sewage, water, or other liquid, partially or completely treated, or in its natural state, as the case may be, flowing out of a reservoir, basin, or treatment plant, or part thereof.

Final.—The effluent from the final unit of a sewage treatment plant.

Stable.—A treated sewage which contains enough oxygen to satisfy its oxygen demand.

Ejector, Pneumatic.—A device for raising sewage, sludge, or other liquid by alternately admitting such through an inward swinging check valve into the bottom of an airtight pot and then discharging it through an outward swinging check valve by admitting compressed air to the pot above the liquid.

Elutriation.—A process of sludge conditioning in which certain constituents are removed by successive decantations with fresh water or plant effluent, thereby reducing the demand for conditioning chemicals.

F

Factor.—Frequently a ratio used to express operating conditions.

Load.—The ratio of the average load carried by any operation to the maximum load carried, during a given period of time, expressed as a percentage. The load may consist of almost anything, such as electrical power, number of persons served, amount of water carried by a conduit, etc.

Power.—An electrical term describing the ratio of the true power passing through an electric circuit to the product of the volts times the amperes in the circuit. It is a measure of the lag or lead of the current in respect to the voltage. While the power of a current is the product of the voltage times the amperes in the circuit, in alternating current the voltage and amperes are not always in phase, hence the true power may be less than that determined by the product of volts times amperes.

Filter.—A term meaning (1) an oxidizing bed (2) a device for removing solids from a liquid by some type of strainer.

Biological.—A bed of sand, gravel, broken stone, or other media through which sewage flows or trickles, which depends on biological action for its effectiveness.

High-Rate.—A trickling filter operated at a high average daily dosing rate usually between 10-30 mgd per acre, sometimes including recirculation of effluent.

Low-Rate.—A trickling filter designed to receive a small load of BOD per unit volume of filtering material and to have a low dosage rate per unit of surface area (usually 1 to 4 mgd per acre). Also called Standard Rate Filter.

Roughing.—A sewage filter of relatively coarse material operated at a high rate as a preliminary treatment.

Sand.—A filter in which sand is used as a filtering medium.

Sand Sludge.—A bed of sand used to dewater sludge by drainage and evaporation.

Sludge.—The solid matter in sewage that is removed by settling in primary and secondary settling tanks.

Trickling.—A treatment unit consisting of a material such as broken stone, clinkers, slate, slats, or brush, over which sewage is distributed and applied in drops, films, or spray, from troughs, drippers, moving distributors, or fixed nozzles, and through which it trickles to the underdrains, giving opportunity for the formation of zoological slimes which clarify and oxidize the sewage.

Vacuum.—A filter consisting of a cylindrical drum mounted on a horizontal axis, covered with filtering material made of wool, felt, cotton, saran, nylon, dacron, polyethylene or similar substance, by stainless steel coil springs or metal screen, revolving with a partial submergence in the liquid. A vacuum is maintained under the cloth for the larger part of a revolution to extract moisture. The cake is scraped off continuously.

Filtrate.—The effluent of a Filter.

Floc.—Small gelatinous masses, formed in a liquid by the addition of coagulants thereto or through biochemical processes or by agglomeration.

Flocculator.—An apparatus for the formation of floc in water or sewage.
Flotation.—A method of raising suspended matter to the surface of the liquid in a tank as scum—by aeration, by the evolution of gas, chemicals, electrolysis, heat, or bacterial decomposition—and the subsequent removal of the scum by skimming.
Freeboard.—The vertical distance between the normal maximum level of the surface of the liquid in a conduit, reservoir, tank, canal, etc., and the top of the sides of an open conduit, the top of a dam or levee, etc., which is provided so that waves and other movements of the liquid will not overtop the confining structure.
Fungi.—Small nonchlorophyll-bearing plants which lack roots, stems, or leaves and which occur (among other places) in water, sewage, or sewage effluents, growing best in the absence of light. Their decomposition after death may cause disagreeable tastes and odors in water; in some sewage treatment processes they are helpful and in others they are detrimental.

G

Gage.—A device for measuring any physical magnitude.
Float.—A device for measuring the elevation of the surface of a liquid, the actuation element being a buoyant float which rests upon the surface of the liquid.
Indicator.—A gage that shows by means of an index, pointer, dial, etc., the instantaneous value of such characteristics as depth, pressure, velocity, stage, discharge, or the movements or positions of water-controlling devices.
Mercury.—A gage wherein pressure of a fluid is measured by the height of a column of mercury which the fluid pressure will sustain. The mercury is usually contained in a tube, attached to the vessel or pipe containing the fluid.
Pressure.—A device for registering the pressure of solids, liquids, or gases. It may be graduated to the register pressure in any units desired.
Garbage, Ground.—Garbage shredded or ground by apparatus installed in sinks and discharged to the sewerage system; or garbage collected and hauled to a central grinding station, shredded preliminary to disposal, usually, by digestion with sewage sludge.
Gas.—One of the three states of matter.
Sewage.—(1) The gas produced by the septicization of sewage. (2) The gas produced during the digestion of sewage sludge, usually collected and utilized.
Sewer.—Gas evolved in sewers from the decomposition of the organic matter in the sewage. Also any gas present in the sewerage system, even though it is from gas mains, gasoline, cleaning fluid, etc.
Gasification.—The transformation of sewage solids into gas in the decomposition of sewage.
Go Devil.—A scraper with self-adjusting spring blades, inserted in a pipe line, and carried forward by the fluid pressure for clearing away accumulations, tuberculations, etc.
Grade.—(1) The inclination or slope of a stream channel, conduit, or natural ground surface, usually expressed in terms of the ratio or percentage of number of units of vertical rise or fall per unit of horizontal distance. (2) The elevation of the invert of the bottom of a pipe line, canal, culvert, sewer, etc. (3) The finished surface of a canal bed, road bed, top of an embankment. or bottom of an excavation. (4) In plumbing, the fall in inches per foot of length of pipe.
Grease.—In sewage, grease including fats, waxes, free fatty acids, calcium and magnesium soaps, mineral oils, and other non-fatty materials. The type of solvent used for its extraction should be stated.

Grinder, Screenings.—A device for grinding, shredding, or comminuting material removed from sewage by screens.

Grit.—The heavy mineral matter in water or sewage, such as gravel, cinders, etc.

H

Head.—Energy per unit weight of liquid at a specified point. It is expressed in feet.

Dynamic.—The head against which a pump works.

Friction.—The head lost by water flowing in a stream or conduit as the result of the disturbances set up by the contact between the moving water and its containing conduit, and by intermolecular friction. In laminar flow the head lost is approximately proportional to the first power of the velocity; in turbulent flow to a higher power, approximately the square of the velocity. While strictly speaking, head losses due to bends, expansions, obstructions, impact, etc., are not included in this term, the usual practice is to include all such head losses under this term.

Loss of.—The decrease in head between two points.

Static.—The vertical distance between the free level of the source of supply, and the point of free discharge, or the level of the free surface.

Total Dynamic.—The difference between the elevation corresponding to the pressure at the discharge flange of a pump and the elevation corresponding to the vacuum or pressure at the suction flange of the pump, corrected to the same datum plane, plus the velocity head at the discharge flange of the pump, minus the velocity head at the suction flange of the pump. It includes the friction head.

Velocity.—The theoretical vertical height through which a liquid body may be raised due to its kinetic energy. It is equal to the square of the velocity divided by twice the acceleration due to gravity.

Humus.—The dark or black carboniferous residue in the soil resulting from the decomposition of vegetable tissues of plants originally growing therein. Residues similar in appearance and behavior are found in well-digested sludges and in activated sludge.

Hypochlorite.—Compounds of chlorine in which the radical (OC1) is present. They are usually inorganic.

High Test.—A solid triple salt containing Ca (0C1) 2 to the extent that the fresh solid has approximately 70 percent available chlorine. It is not the same as chlorinated lime.

Sodium.—A solution containing Na0C1, prepared by passing chlorine into solutions of soda ash, or reacting soda ash solutions with high-test hypochlorites and decanting from the precipitated sludge.

I

Imhoff Cone.—A conically shaped graduated glass vessel used to measure approximately the volume of settleable solids in various liquids of sewage origin.

Imhoff Tank.—See Tank, Imhoff

Impeller.—The rotating part of a centrifugal pump, containing the curved vanes.

Closed.—An impeller having the side walls extended from the outer circumference of the suction opening to the vane tips.

Nonclogging.—An impeller of the open, closed, or semi-closed type designed with large passages for passing large solids.

Open.—An impeller without attached side walls.

Screw.—The helical impeller of a screw pump.

Index, Sludge Volume.—The volume is milliliters occupied by one gram of dry solids after the aerated mixed liquor settles 30 minutes, commonly referred to as the Mohlman index.

Influent.—Sewage, water, or other liquid, raw or partly treated, flowing into a reservoir, basin, or treatment plant, or part thereof.

L

Lagoon, Sludge.—A relatively shallow basin, or natural depression, used for the storage or digestion of sludge, and sometimes for its ultimate detention or dewatering.

Lift, Air.—A device for raising liquid by injecting air in and near the bottom of a riser pipe submerged in the liquid to be raised.

Liquefaction.—The changing of the organic matter in sewage from an insoluble to a soluble state, and effecting a reduction in its solid contents.

Liquor.—Any liquid.

Mixed.—A mixture of activated sludge and sewage in the aeration tank undergoing activated sludge treatment.

Supernatant. — (1) The liquor overlying deposited solids. (2) The liquid in a sludge-digestion tank which lies between the sludge at the bottom and the floating scum at the top.

Loading.—The time rate at which material is applied to a treatment device involving length, area, or volume or other design factor.

BOD, Filter.—The pounds of oxygen demand in the applied liquid per unit of filter bed area, or volume of stone per day.

Weir.—Gallons overflow per day per foot of weir length.

M

Main, Force.—A pipe line on the discharge side of a water or sewage pumping station, usually under pressure.

Manometer.—An instrument for measuring pressure; usually it consists of a U-shaped tube containing a liquid, the surface of which in one end of the tube moves proportionally with changes in pressure upon the liquid in the other end. The term is also applied to a tube type of differential pressure gage.

Matter.—Solids, liquids, and gases.

Inorganic.—Chemical substances of mineral origin. They are not usually volatile with heat.

Organic.—Chemical substances of animal, vegetable and industrial origin. They include most carbon compounds, combustible and volatile with heat.

Suspended.—(1) Solids in suspension in sewage or effluent. (2) Commonly used for solids in suspension in sewage or effluent which can readily be removed by filtering in a laboratory.

Microorganism.—Minute organisms either plant or animal, invisible or barely visible to the naked eye.

Moisture, Percentage.—The water content of sludge expressed as the ratio of the loss in weight after drying at 103°C, to the original weight of the sample, multiplied by one hundred.

Mold.—See Fungi.

Most Probable Number, (MPN).—In the testing of bacterial density by the dilution method, that number of organisms per unit volume which, in accordance with statistical theory, would be more likely than any other possible number to yield the observed test result or which would yield the observed test result with the greatest frequency. Expressed as density of organisms per 100 ml.

N

Nitrification.—The oxidation of ammonia nitrogen into nitrates through biochemical action.

O

Overflow Rate.—One of the criteria for the design of settling tanks in treatment plants; expressed in gallons per day per square foot of surface area in the settling tank. See Surface Settling Rate.

Oxidation.—The addition of oxygen, removal of hydrogen, or the increase in the valence of an element.

- **Biochemical.**—See Oxidation, Sewage.
- **Biological.**—See Oxidation, Sewage.
- **Direct.**—Oxidation of substances in sewage without the benefit of living organisms, by the direct application of air or oxidizing agents such as chlorine.
- **Sewage.**—The process whereby, through the agency of living organisms in the presence of oxygen, the organic matter contained in sewage is converted into a more stable form.

Oxygen.—A chemical element.

- **Available.**—The quantity of uncombined or free oxygen dissolved in the water of a stream.
- **Balance.**—The relation between the biochemical oxygen demand of a sewage or treatment plant effluent and the oxygen available in the diluting water.
- **Consumed.**—The quantity of oxygen taken from potassium permanganate in solution by a liquid containing organic matter. Commonly regarded as an index of the carbonaceous matter present. Time and temperature must be specified. The chemical oxygen demand (COD) uses potassium dichromate.
- **Deficiency.**—The additional quantity of oxygen required to satisfy the biochemical oxygen demand in a given liquid. Usually expressed in parts per million.
- **Dissolved.**—Usually designated as DO. The oxygen dissolved in sewage, water or other liquid usually expressed in parts per million or percent of saturation.
- **Residual.**—The dissolved oxygen content of a stream after deoxygenation has begun.
- **Sag.**—A curve that represents the profile of dissolved oxygen content along the course of a stream, resulting from deoxygenation associated with biochemical oxidation of organic matter, and reoxygenation through the absorption of atmospheric oxygen and through biological photosynthesis.

P

Parts Per Million.—Milligrams per liter expressing the concentration of a specified component in a dilute sewage. A ratio of pounds per million pounds, grams per million grams, etc.

Percolation.—The flow or trickling of a liquid downward through a contact or filtering medium. The liquid may or may not fill the pores of the medium.

Period.—A time interval.

Aeration.—(1) The theoretical time, usually expressed in hours that the mixed liquor is subjected to aeration in an aeration tank undergoing activated sludge treatment; is equal to (a) the volume of the tank divided by (b) the volumetric rate of flow of the sewage and return sludge. (2) The theoretical time that water is subjected to aeration.

Detention.—The theoretical time required to displace the contents of a tank or unit at a given rate of discharge (volume divided by rate of discharge).

Flowing-Through.—The average time required for a small unit volume of liquid to pass through a basin from inlet to outlet. In a tank where there is no short-circuiting, and no spaces, the detention period and the flowing-through period are the same.

pH.—The logarithm of the reciprocal of the hydrogen-ion concentration. It is not the same as the alkalinity and cannot be calculated therefrom.

Plankton.—Drifting organisms, usually microscopic.

Pollution.—The addition of sewage, industrial wastes, or other harmful or objectionable material to water.

Ponding, Filter.—See Pooling, Filter.

Pooling, Filter.—The formation of pools of sewage on the surface of filters caused by clogging.

Population Equivalent.—(1) The calculated population which would normally contribute the same amount of biochemical oxygen demand (BOD) per day. A common base is 0.167 lb. of 5-day BOD per capita per day. (2) For an industrial waste, the estimated number of people contributing sewage equal in strength to a unit volume of the waste or to some other unit involved in producing or manufacturing a particular commodity.

Pre-Aeration.—A preparatory treatment of sewage comprising aeration to remove gases, add oxygen, or promote flotation of grease, and aid coagulation.

Precipitation, Chemical.—Precipitation induced by addition of chemicals.

Pressure.—Pounds per square inch or square foot.

Atmospheric.—The pressure exerted by the atmosphere at any point. Such pressure decreases the elevation of the point above sea level increases. One atmosphere is equal to 14.7 lb. per sq. in., 29.92 in. or 760 mm of mercury column or 33.90 ft. of water column at average sea level under standard conditions.

Hydrostatic.—The pressure, expressed as a total force per unit of area, exerted by a body of water at rest.

Negative.—A pressure less than the local atmospheric pressure at a given point.

Process.—A sequence of operations.

Activated Sludge.—A biological sewage treatment process in which a mixture of sewage and activated sludge is agitated and aerated. The activated sludge is subsequently separated from the treated sewage (mixed liquor) by sedimentation, and wasted or returned to the process as needed. The treated sewage overflows the weir of the settling tank in which separation from the sludge takes place.

Biological.—The process by which the life activities of bacteria, and other microorganisms in the search for food, break down complex organic materials into simple, more stable substances. Self-purification of sewage-polluted streams, sludge digestion, and all so-called secondary sewage treatments result from this process. Also called Biochemical Process.

Pump.--A device used to increase the head on a liquid.

Booster.—A pump installed on a pipe line to raise the pressure of the water on the discharge side of the pump.

Centrifugal, Fluid.—A pump consisting of an impeller fixed on a rotating shaft and enclosed in a casing, having an inlet and a discharge connection. The rotating impeller creates pressure in the liquid by the velocity derived from centrifugal force.

Centrifugal, Screw.—A centrifugal pump having a screw-type impeller; may be axial-flow, or combined axial and radial-flow, type.

Centrifugal, Closed.—A centrifugal pump where the impeller is built with the vanes enclosed within circular disks.

Diaphragm.—A pump in which a flexible diaphragm, generally of rubber, is the operating part; it is fastened at the outer rim; when the diaphragm is moved in one direction, suction is exerted and when it is moved in the opposite direction, the liquid is forced through a discharge valve.

Double-Suction.—A centrifugal pump with suction pipes connected to the casing from both sides.

Duplex.—A reciprocating pump consisting of two cylinders placed side by side and connected to the same suction and discharge pipe, the pistons moving so that one exerts suction while the other exerts pressure, with the result that the discharge from the pump is continuous.

Horizontal Screw.—A pump with a horizontal cylindrical casing, in which operates a runner with radial blades, like those of a ship's propeller. The pump has a high efficiency at low heads and high discharges, and is used extensively in drainage work.

Mixed Flow.—A centrifugal pump in which the head is developed partly by centrifugal force and partly by the lift of the vanes on the liquid.

Open Centrifugal.—A centrifugal pump where the impeller is built with a set of independent vanes.

Propeller.—A centrifugal pump which develops most of its head by the propelling or lifting action of the vanes on the liquids.

Purification.—The removal, by natural or artificial methods, or objectionable matter from water.

Putrefaction.—Biological decomposition of organic matter with the production of ill-smelling products associated with anaerobic conditions.

Putrescibility. — (1) The relative tendency of organic matter to undergo decomposition in the absence of oxygen. (2) The susceptibility of waste waters, sewage, effluent, or sludge to putrefaction. (3) Term used in water or sewage analysis to define stability of a polluted water or raw or partially treated sewage.

Q

Quicklime.—A calcined material, the major part of which is calcium oxide or calcium oxide in natural association with a lesser amount of magnesium oxide, capable of slaking with water.

R

Rack.—An arrangement of parallel bars.

Bar.—A screen composed of parallel bars, either vertical or inclined, placed in a waterway to catch floating debris, and from which the screenings may be raked. Also called rack.

Coarse.—A rack with 3/4 inch to 6 inch spaces between bars.

Fine.—Generally used for a screen or rack which has openings of 3/32 to 3/16 inches. Some screens have less than 3/32 inch openings.

Radius, Hydraulic.—The cross-sectional area of a stream of water divided by the length of that part of its periphery in contact with its containing conduit; the ratio of area to wetted perimeter.

Rate.—The result of dividing one concrete number by another.

Filtration.—The rate of application of water or sewage to a filter, usually expressed in million gallons per acre per day, or gallons per minute per square foot.

Infiltration.—The rate, usually expressed in cubic feet per second, or million gallons per day per mile of waterway, at which ground water enters an infiltration ditch or gallery, drain, sewer, or other underground conduit.

Surface Settling.—Gallons per day per square foot of free horizontal water surface. Used in design of sedimentation tanks.

Reaeration.—The absorption of oxygen by a liquid, the dissolved oxygen content of which has been depleted.

Reaeration, Sludge.—The continuous aeration of sludge after its initial aeration in the activated sludge process.

Recirculation. — (1) The refiltration of all or a portion of the effluent in a high-rate trickling filter for the purpose of maintaining a uniform high rate through the filter. (2) The return of effluent to the incoming flow to reduce its strength.

Reduction.—The decrease in a specific variable.

Over-All.—The percentage reduction in the final effluent as compared to the raw sewage.

Percentage.—The ratio of material removed from water or sewage by treatment, to the material originally present (expressed as a percentage).

Sludge.—The reduction in the quantity and change in character of sewage sludge as the result of digestion.

Regulator.—A device or apparatus for controlling the quantity of sewage admitted to an intercepting sewer or a unit of a sewage treatment plant.

Reoxygenation.—The replenishment of oxygen in a stream from (1) dilution water entering stream, (2) biological reoxygenation through the activities of certain oxygen-producing plants, and (3) atmospheric reaction.

Residual, Chlorine.—See Chlorine, residual.

Rotor.—The member of an electric generator or water wheel which rotates.

S

Screen.—A device with openings, generally of uniform size, used to retain or remove suspended or floating solids in flowing water or sewage, and to prevent them from entering an intake or passing a given point in a conduit. The screening element may consist of parallel bars, rods, wires, grating, wire mesh, or perforated plate, and the openings may be of any shape, although they are generally circular or rectangular. The device may also be used to segregate granular material, such as sand, crushed rock, and soil, into various sizes.

Scum.—A mass of sewage matter which floats on the surface of sewage.

Second-Foot.—An abbreviated expression for cubic foot per second.

Sedimentation.—The process of subsidence and deposition of suspended matter carried by water, sewage, or other liquids, by gravity. It is usually accomplished by reducing the velocity of the liquid below the point where it can transport the suspended material. Also called Settling. See Precipitation, Chemical.

Final.—Settling of partly settled, flocculated or oxidized sewage in a final tank.

Plain.—The sedimentation of suspended matter in a liquid unaided by chemicals or other special means, and without provision for the decomposition of deposited solids in contact with the sewage.

Seeding, Sludge.—The inoculation of undigested sewage solids with sludge that has undergone decomposition, for the purpose of introducing favorable organisms, thereby accelerating the initial stages of digestion.

Self-Purification.—The natural processes of purification in a moving or still body of water whereby the bacterial content is reduced, the BOD is largely satisfied, the organic content is stabilized, and the dissolved oxygen returned to normal.

Sewage.—Largely the water supply of a community after it has been fouled by various uses. From the standpoint of source it may be a combination of the liquid or water-carried wastes from residences, business buildings, and institutions, together with those from industrial establishments, and with such ground water, surface water, and storm water as may be present.

Domestic.—Sewage derived principally from dwellings, business buildings, institutions, and the like. (It may or may not contain ground water, surface water, or storm water.)

Fresh.—Sewage of recent origin containing dissolved oxygen at the point of examination.

Industrial.—Sewage in which industrial wastes predominate.

Stable.—Sewage in which the organic matter has been stabilized.

Raw.—Sewage prior to receiving any treatment.

Sanitary.—(1) Domestic sewage with storm and surface water excluded. (2) Sewage discharging from the sanitary conveniences of dwellings (including apartment houses and hotels), office buildings, factories, or institutions. (3) The water supply of a community after it has been used and discharged into a sewer.

Septic.—Sewage undergoing putrefaction under anaerobic conditions.

Settled.—Sewage from which most of the settleable solids have been removed by sedimentation.

Stale.—A sewage containing little or no oxygen, but as yet free from putrefaction.

Sewer.—A pipe or conduit. generally closed, but normally not flowing full, for carrying sewage and other waste liquids.

Branch.—A sewer which receives sewage from a relatively small area, and discharges into a main sewer.

Combined.—A sewer receiving both surface runoff and sewage.

House.—A pipe conveying sewage from a single building to a common sewer or point of immediate disposal.

Intercepting.—A sewer which receives dry-weather flow from a number of transverse sewers or outlets and frequently additional predetermined quantities of storm water (if from a combined system), and conducts such waters to a point for treatment or disposal.

Lateral.—A sewer which discharges into a branch or other sewer and has no other common sewer tributary to it.

Main. — (1) A sewer to which one or more branch sewers are tributary. Also called Trunk Sewer. (2) In plumbing, the public sewer in a street, alley, or other premises under the jurisdiction of a municipality.

Sanitary.—A sewer which carries sewage and to which storm, surface, and ground waters are not intentionally admitted.

Separate.—See Sewer, Sanitary.

Storm.—A sewer which carries storm water and surface water, street wash and other wash waters, or drainage, but excludes sewage and industrial wastes. Also called Storm Drain.

Trunk—A sewer which receives many tributary branches and serves a large territory. See Sewer, Main.

Outfall.—A sewer which receives the sewage from a collecting system and carries it to a point of final discharge.

Outlet.—The point of final discharge of sewage or treatment plant effluent.

Sewerage.—A comprehensive term which includes facilities for collecting, pumping, treating, and disposing of sewage; the sewerage system and the sewage treatment works.

Shredder.—A device for size reduction.

Screenings.—A device which disintegrates screenings.

Sludge.—An apparatus to break down lumps in air-dried digested sludge.

Siphon.—A closed conduit, a portion of which lies above the hydraulic grade line. This results in a pressure less than atmospheric in that portion, and hence requires that a vacuum be created to start flow.

Skimmer, Grease.—A device for removing floating grease or scum from the surface of sewage in a tank.

Skimming.—The process of removing floating grease or scum from the surface of sewage in a tank.

Sleek.—The thin oily film usually present which gives characteristic appearance to the surface of water into which sewage or oily waste has discharged. Also termed slick.

Sloughing.—The phenomenon associated with trickling filters and contact aerators. whereby slime and solids accumulated in the media are discharged with the effluent.

Sludge.—The accumulated settled solids deposited from sewage or industrial wastes, raw or treated. in tanks or basins, and containing more or less water to form a semiliquid mass.

Activated.—Sludge floc produced in raw or settled sewage by the growth of zoogleal bacteria and other organisms in the presence of dissolved oxygen, and accumulated in sufficient concentration by returning floc previously formed.

Bulking.—A phenomenon that occurs in activated sludge plants whereby the sludge occupies excessive volumes and will not concentrate readily.

Conditioning.—Treatment of liquid sludge preliminary to dewatering and drainability, usually by the addition of chemicals.

Dewatering.—The process of removing a part of the water in sludge by any method, such as draining, evaporation, pressing, centrifuging, exhausting, passing between rollers, or acid flotation, with or without heat. It involves reducing from a liquid to a spadable condition rather than merely changing the density of the liquid (concentration) on the one hand or drying (as in a kiln) on the other.

Digestion.—The process by which organic or volatile matter in sludge is gasified, liquefied, mineralized. or converted into more stable organic matter, through the activities of living organisms.

Humus.—See **Humus.**

Solids.—Material in the solid state.

Dissolved.—Solids which are present in solution.

Nonsettleable.—Finely divided suspended solids which will not subside in quiescent water, sewage, or other liquid in a reasonable period. Such period is commonly, though arbitrarily, taken as two hours.

Settleable.—Suspended solids which will subside in quiescent water, sewage, or other liquid in a reasonable period. Such period is commonly, though arbitrarily, taken as one hour. Also called Settling Solids.

Suspended.—The quantity of material deposited when a quantity of water, sewage, or other liquid is filtered through an asbestos mat in a Gooch crucible.

Total.—The solids in water, sewage, or other liquids; it includes the suspended solids (largely removable by filter paper) and the filterable solids (those which pass through filter paper).

Volatile.—The quantity of solids in water, sewage, or other liquid, lost on ignition of the total solids.

Squeegee.—(1) A device, generally with a soft rubber edge, used for dislodging and removing deposited sewage solids from the walls and bottoms of sedimentation tanks. (2) The metal blades attached to the lower arms of a clarifier mechanism to move the sludge along the tank bottom.

Stability.—The ability of any substance, such as sewage, effluent, or digested sludge, to resist putrefaction. It is the antonym of putrescibility.

Standard Methods.—Methods of analysis of water, sewage, and sludge approved by a Joint Committee of the American Public Health Association. American Water Works Association, and Federation of Sewage Works Associations.

Stator.—The stationary member of an electric generator or motor.

Sterilization.—The destruction of all living organisms, ordinarily through the agency of heat or of some chemical.

T

Tank.—A circular or rectangular vessel.

Detritus.—A detention chamber larger than a grit chamber, usually with provision for removing the sediment without interrupting the flow of sewage. A settling tank of short detention period designed, primarily, to remove heavy settleable solids.

Final Settling.—A tank through which the effluent from a trickling filter, or aeration or contact aeration tank flows for the purpose of removing the settleable solids.

Flocculating.—A tank used for the formation of floc by the agitation of liquids.

Imhoff.—A deep two-storied sewage tank originally patented by Karl Imhoff. consisting of an upper or continuous flow sedimentation chamber and a lower or sludge-digestion chamber. The floor of the upper chamber slopes steeply to trapped slots, through which solids may slide into the lower chamber. The lower chamber receives no fresh sewage directly, but is provided with gas vents and with means for drawing digested sludge from near the bottom.

Primary Settling.—The first settling tank through which sewage is passed in a treatment works.

Secondary.—A tank following a trickling filter or activated sludge aeration chamber.

Sedimentation.—A tank or basin. in which water, sewage, or other liquid containing settleable solids, is retained for a sufficient time, and in which the velocity of flow is sufficiently low, to remove by gravity a part of the suspended matter. Usually, in sewage treatment, the detention period is short enough to avoid anaerobic decomposition. Also termed Settling or Subsidence Tank.

Septic.—A single-story settling tank in which the settled sludge is in immediate contact with the sewage flowing through the tank, while the organic solids are decomposed by anaerobic bacterial action.

Sludge-Digestion--See Digester.

Thickener, Sludge.—A type of sedimentation tank in which the sludge is permitted to settle, usually equipped with scrapers traveling along the bottom of the tank which push the settled sludge to a sump, from which it is removed by gravity or by pumping.

Treatment.—Any definite process for modifying the state of matter.

Preliminary.—The conditioning of an industrial waste at its source prior 'to discharge, to remove or to neutralize substances injurious to sewers and treatment processes or to effect a partial reduction in load on the treatment process. In the treatment process, unit operations which prepare the liquor for subsequent major operations.

Primary.—The first major (sometimes the only) treatment in a sewage treatment works, usually sedimentation. The removal of a high percentage of suspended matter but little or no colloidal and dissolved matter.

Secondary.—The treatment of sewage by biological methods after primary treatment by sedimentation.

Sewage.—Any artificial process to which sewage is subjected in order to remove or alter its objectional constituents and thus to render it less offensive or dangerous.

Trap, Flame.—A device containing a fine metal gauze placed in a gas pipe, which prevents a flame from traveling back in the pipe and causing an explosion. See Arrester, Flame.

V

Venturi Meter.—A meter for measuring flow of water or other fluid through closed conduits or pipes, consisting of a Venturi tube and one of several proprietary forms of flow registering devices. The device was developed as a measuring device and patented by Clemens Herschel.

W

Waste Stabilization Pond.—Any pond, natural or artificial, receiving raw or partially treated sewage or waste, in which stabilization occurs due to sunlight, air, and microorganisms.

Water, Potable.—Water which does not contain objectionable pollution, contamination, minerals, or infection, and is considered satisfactory for domestic consumption.

Weir.—A dam with an edge or notch, sometimes arranged for measuring liquid flow.

Effluent —A weir at the outflow end of a sedimentation basin or other hydraulic structure.

Influent —A weir at the inflow end of a sedimentation basin.

Rectangular.—A weir whose notch is rectangular in shape.

Triangular.—A weir whose notch is triangular in shape, usually used to measure very small flows. Also called a V-notch.

Peripheral.—The outlet weir in a circular settling tank, extending around the inside of its circumference and over which the effluent discharges.

Rate.—See Loading, Weir.

Z

Zooglea.—A jelly-like matrix developed by bacteria, associated with growths in oxidizing beds.

ANSWER SHEET

TEST NO. ______________ PART ______ TITLE OF POSITION ______________________________
(AS GIVEN IN EXAMINATION ANNOUNCEMENT - INCLUDE OPTION, IF ANY)

PLACE OF EXAMINATION ______________________________ ______________ DATE ______________
(CITY OR TOWN) (STATE)

RATING

USE THE SPECIAL PENCIL. MAKE GLOSSY BLACK MARKS.

No.	A	B	C	D	E	No.	A	B	C	D	E	No.	A	B	C	D	E	No.	A	B	C	D	E	No.	A	B	C	D	E
1	A	B	C	D	E	26	A	B	C	D	E	51	A	B	C	D	E	76	A	B	C	D	E	101	A	B	C	D	E
2	A	B	C	D	E	27	A	B	C	D	E	52	A	B	C	D	E	77	A	B	C	D	E	102	A	B	C	D	E
3	A	B	C	D	E	28	A	B	C	D	E	53	A	B	C	D	E	78	A	B	C	D	E	103	A	B	C	D	E
4	A	B	C	D	E	29	A	B	C	D	E	54	A	B	C	D	E	79	A	B	C	D	E	104	A	B	C	D	E
5	A	B	C	D	E	30	A	B	C	D	E	55	A	B	C	D	E	80	A	B	C	D	E	105	A	B	C	D	E
6	A	B	C	D	E	31	A	B	C	D	E	56	A	B	C	D	E	81	A	B	C	D	E	106	A	B	C	D	E
7	A	B	C	D	E	32	A	B	C	D	E	57	A	B	C	D	E	82	A	B	C	D	E	107	A	B	C	D	E
8	A	B	C	D	E	33	A	B	C	D	E	58	A	B	C	D	E	83	A	B	C	D	E	108	A	B	C	D	E
9	A	B	C	D	E	34	A	B	C	D	E	59	A	B	C	D	E	84	A	B	C	D	E	109	A	B	C	D	E
10	A	B	C	D	E	35	A	B	C	D	E	60	A	B	C	D	E	85	A	B	C	D	E	110	A	B	C	D	E

Make only ONE mark for each answer. Additional and stray marks may be counted as mistakes. In making corrections, erase errors COMPLETELY.

No.	A	B	C	D	E	No.	A	B	C	D	E	No.	A	B	C	D	E	No.	A	B	C	D	E	No.	A	B	C	D	E
11	A	B	C	D	E	36	A	B	C	D	E	61	A	B	C	D	E	86	A	B	C	D	E	111	A	B	C	D	E
12	A	B	C	D	E	37	A	B	C	D	E	62	A	B	C	D	E	87	A	B	C	D	E	112	A	B	C	D	E
13	A	B	C	D	E	38	A	B	C	D	E	63	A	B	C	D	E	88	A	B	C	D	E	113	A	B	C	D	E
14	A	B	C	D	E	39	A	B	C	D	E	64	A	B	C	D	E	89	A	B	C	D	E	114	A	B	C	D	E
15	A	B	C	D	E	40	A	B	C	D	E	65	A	B	C	D	E	90	A	B	C	D	E	115	A	B	C	D	E
16	A	B	C	D	E	41	A	B	C	D	E	66	A	B	C	D	E	91	A	B	C	D	E	116	A	B	C	D	E
17	A	B	C	D	E	42	A	B	C	D	E	67	A	B	C	D	E	92	A	B	C	D	E	117	A	B	C	D	E
18	A	B	C	D	E	43	A	B	C	D	E	68	A	B	C	D	E	93	A	B	C	D	E	118	A	B	C	D	E
19	A	B	C	D	E	44	A	B	C	D	E	69	A	B	C	D	E	94	A	B	C	D	E	119	A	B	C	D	E
20	A	B	C	D	E	45	A	B	C	D	E	70	A	B	C	D	E	95	A	B	C	D	E	120	A	B	C	D	E
21	A	B	C	D	E	46	A	B	C	D	E	71	A	B	C	D	E	96	A	B	C	D	E	121	A	B	C	D	E
22	A	B	C	D	E	47	A	B	C	D	E	72	A	B	C	D	E	97	A	B	C		E	122	A	B	C	D	E
23	A	B	C	D	E	48	A	B	C	D	E	73	A	B	C	D	E	98	A	B	C	D	E	123	A	B	C	D	E
24	A	B	C	D	E	49	A	B	C	D	E	74	A	B	C	D	E	99	A	B	C	D	E	124	A	B	C	D	E
25	A	B	C	D	E	50	A	B	C	D	E	75	A	B	C	D	E	100	A	B	C	D	E	125	A	B	C	D	E

ANSWER SHEET

TEST NO. ________________ PART ______ TITLE OF POSITION ______________________________
(AS GIVEN IN EXAMINATION ANNOUNCEMENT - INCLUDE OPTION, IF ANY)

PLACE OF EXAMINATION ______________________________ ______________________ DATE ______________
(CITY OR TOWN) (STATE)

RATING

USE THE SPECIAL PENCIL. MAKE GLOSSY BLACK MARKS.

1	A B C D E	26	A B C D E	51	A B C D E	76	A B C D E	101	A B C D E
2	A B C D E	27	A B C D E	52	A B C D E	77	A B C D E	102	A B C D E
3	A B C D E	28	A B C D E	53	A B C D E	78	A B C D E	103	A B C D E
4	A B C D E	29	A B C D E	54	A B C D E	79	A B C D E	104	A B C D E
5	A B C D E	30	A B C D E	55	A B C D E	80	A B C D E	105	A B C D E
6	A B C D E	31	A B C D E	56	A B C D E	81	A B C D E	106	A B C D E
7	A B C D E	32	A B C D E	57	A B C D E	82	A B C D E	107	A B C D E
8	A B C D E	33	A B C D E	58	A B C D E	83	A B C D E	108	A B C D E
9	A B C D E	34	A B C D E	59	A B C D E	84	A B C D E	109	A B C D E
10	A B C D E	35	A B C D E	60	A B C D E	85	A B C D E	110	A B C D E

Make only ONE mark for each answer. Additional and stray marks may be counted as mistakes. In making corrections, erase errors COMPLETELY.

11	A B C D E	36	A B C D E	61	A B C D E	86	A B C D E	111	A B C D E
12	A B C D E	37	A B C D E	62	A B C D E	87	A B C D E	112	A B C D E
13	A B C D E	38	A B C D E	63	A B C D E	88	A B C D E	113	A B C D E
14	A B C D E	39	A B C D E	64	A B C D E	89	A B C D E	114	A B C D E
15	A B C D E	40	A B C D E	65	A B C D E	90	A B C D E	115	A B C D E
16	A B C D E	41	A B C D E	66	A B C D E	91	A B C D E	116	A B C D E
17	A B C D E	42	A B C D E	67	A B C D E	92	A B C D E	117	A B C D E
18	A B C D E	43	A B C D E	68	A B C D E	93	A B C D E	118	A B C D E
19	A B C D E	44	A B C D E	69	A B C D E	94	A B C D E	119	A B C D E
20	A B C D E	45	A B C D E	70	A B C D E	95	A B C D E	120	A B C D E
21	A B C D E	46	A B C D E	71	A B C D E	96	A B C D E	121	A B C D E
22	A B C D E	47	A B C D E	72	A B C D E	97	A B C D E	122	A B C D E
23	A B C D E	48	A B C D E	73	A B C D E	98	A B C D E	123	A B C D E
24	A B C D E	49	A B C D E	74	A B C D E	99	A B C D E	124	A B C D E
25	A B C D E	50	A B C D E	75	A B C D E	100	A B C D E	125	A B C D E